Synthesis Lectures on Engin
and Technology

The focus of this series is general topics, and applications about, and for, engineers and scientists on a wide array of applications, methods and advances. Most titles cover subjects such as professional development, education, and study skills, as well as basic introductory undergraduate material and other topics appropriate for a broader and less technical audience.

Arturo Realyvásquez Vargas ·
Jorge Luis García Alcaraz ·
Suchismita Satapathy · José Roberto Díaz-Reza

The PDCA Cycle for Industrial Improvement

Applied Case Studies

Arturo Realyvásquez Vargas
Department of Industrial Engineering
Tecnológico Nacional de México/IT Tijuana
Tijuana, Baja California, Mexico

Suchismita Satapathy
School of Mechanical Engineering
Kalinga Institute of Industrial Technology
Bhubaneswar, Odisha, India

Jorge Luis García Alcaraz
Department of Industrial Engineering
and Manufacturing
Universidad Autónoma de Ciudad Juárez
Ciudad Juárez, Chihuahua, Mexico

José Roberto Díaz-Reza
Department of Electric Engineering
and Computer Sciences
Universidad Autónoma de Ciudad Juárez
Ciudad Juárez, Chihuahua, Mexico

ISSN 2690-0300 ISSN 2690-0327 (electronic)
Synthesis Lectures on Engineering, Science, and Technology
ISBN 978-3-031-26807-6 ISBN 978-3-031-26805-2 (eBook)
https://doi.org/10.1007/978-3-031-26805-2

This Springer imprint is published by the registered company Springer Nature Switzerland AG
The registered company address is: Gewerbestrasse 11, 6330 Cham, Switzerland

Preface

Lean Manufacturing (LM) is a production philosophy that integrates tools focused on eliminating waste. Traditionally, LM is represented by a house that integrates basic tools and pillars that support the flow of materials, quality assurance, manufacturing systems, and human resources.

The basic LM tools are focused on the standardization of the production system, among which are the visual factory, 5S, Kaizen, total productive maintenance and six sigma, to mention just a few. The pillar associated with production flow are tools such as SMED, Kan-ban, push systems, and just-in-time, among others; the pillar associated with quality assurance is made up of tools such as poka-yoke, quality at the source, value stream mapping, total quality management, among others, while the tools associated with manufacturing systems are made up of Layout, batch production, cellular manufacturing, multifunctional work areas, among others.

A tool considered the origin of many others refers to the Plan-Do-Check-Act (PDCA) cycle, which is used for problem-solving and improvement in production systems. This tool has proven its efficiency over the years in different industrial sectors due to its simple and easy application, and unfortunately, it is often ignored or under-appreciated. This book aims to provide a conceptual description of the PDCA cycle and the tools that support it and to report some examples. The book consists of four chapters, which are described below.

Chapter 1 gives some definitions and concepts about the PDCA tool, as well as those that support it, where an example of application is given for each one. Additionally, a brief literature review is given for each tool, indicating the timeline according to Scopus, the main areas in which they are applied, the main authors and the journals that most publish on these topics. In this case, the auxiliary tools are the flow chart, Pareto diagram, process flow chart, Ishikawa diagram and 5s.

Chapter 2 reports the first case study in which the PDCA cycle is applied and focuses on cost reduction in disposing of glass and general waste. The results indicate a reduction of 48.45% of the cost of general waste and 70.59% of fiberglass waste. In addition, 15% of the waste was reprocessed or marketed to other companies.

Chapter 3 reports the second case study in which the PDCA cycle is applied and focuses on optimizing a manufacturing company's raw material receiving process. Initially, the process took up to 45 days due to storage space problems and incurred a cost of US$50,000 per year in rent payments. Applying the PDCA cycle reduced the lead time to just 7 days and increased the efficiency of certain operations. For example, labeling went from 580 units to 1,470 units, and 17 activities that did not add value to the product or were combined with another activity were eliminated.

Finally, Chap. 4 reports the third case study in which the PDCA cycle is applied in a company that manufactures high-voltage electrical cables. The application focuses on reducing waste associated with lead times, excess transportation, overproduction, excess defects in finished products and financial underperformance. After applying the PDCA cycle, the production rate was increased from 432 to 813 units produced per shift, which increased the company's financial performance.

Tijuana, Mexico Arturo Realyvásquez Vargas
Ciudad Juárez, Mexico Jorge Luis García Alcaraz
Bhubaneswar, India Suchismita Satapathy
Ciudad Juárez, Mexico José Roberto Díaz-Reza

Contents

List of Figures

List of Tables

Plan-Do-Check-Act Cycle (PDCA) and Auxiliary Tools for Troubleshooting Manufacturing Processes

1

1.1 PDCA Cycle Concept

The Plan-Do-Check-Act (PDCA) cycle, also known as the Deming cycle [1], is a lean manufacturing tool [2, 3]; more specifically, it is a quality management system applied in various sectors such as manufacturing, services, offshore, project areas, organizations, among others [4]. The PDCA cycle was popularized by Edwards Deming, an American expert in quality management, in 1950.

In its early days, PDCA was used as a tool for product quality control [3, 5], and currently, it is one of the most well-known and widely used methods through which quality tools can be applied in the operations of a manufacturing process [6]. This method is useful for making continuous, incremental, rapid and effective improvements without large capital investments at the organizational level; as compared to other methods, it is less complicated and costly [3–9].

The PDCA cycle consists of four phases: Plan, Do, Check, and Act [10, 11], which are mentioned continued below [12]:

Plan: This phase consists of identifying the problem, and for this purpose, the problem to be analyzed is selected, and a precise statement of the problem is established. In addition, measurable objectives for the problem-solving effort are established, and a process for coordinating and obtaining leadership approval.

Next, the problem is analyzed, the processes that impact the problem are identified, and one is selected. Next, the process activities are listed as they occur, the potential root cause of the problem is identified, and the data related to the problem is collected and analyzed. Finally, the original problem statement is reviewed and verified, and root causes are identified.

Do: This phase consists of developing and implementing solutions. Therefore, first, criteria for selecting a solution must be established. Next, potential solutions that address

A. Realyvásquez Vargas et al., *The PDCA Cycle for Industrial Improvement*, Synthesis Lectures on Engineering, Science, and Technology,
https://doi.org/10.1007/978-3-031-26805-2_1

the root causes of the problem are generated, and finally, a solution is selected, and a plan is developed to implement it on a trial or pilot basis.

Check: In this phase, the results are evaluated by gathering data obtained after implementing the solution and analyzing them.

Act: In this phase, the next steps are determined. If the desired objective is not achieved, the PDCA cycle is repeated. If the objective is achieved, system changes necessary for a complete solution implementation are identified. Finally, the solution is adopted, the results are monitored, and the next opportunity for improvement is sought.

To apply the different phases of the PDCA cycle, one can resort to the use of some quality tools [3], such as the flow chart, the Pareto diagram, the flow process chart, Ishikawa diagram, and the 5's [3, 13], which are detailed in the following sections.

1.1.1 PDCA—Brief Literature Review

The applications of the PDCA tool have been widely reported in the literature over time and in different areas of knowledge. A review has been carried out in Scopus with the following search equation: (TITLE-ABS-KEY (PDCA) AND TITLE-ABS-KEY (quality OR improvement)) and 969 documents have been found.

Figure 1.1 illustrates the evolution of the use of the PDCA tool associated with quality and process improvement. The blue line indicates the number of documents found, and the red dotted line represents a quadratic trend line with a 95.28% fit. From the blue line, it can be deduced that this combination of words appears for the first time in 1981, and after almost a decade, it is not until 1989 that they appear again with four documents, and from then on, the number of documents increases.

Figure 1.1 shows that it is from 1993 onwards that the application of the PDCA cycle in the industry began to be documented and its case studies reported, even though its origins date back to the 1950s. For the year 2021, for example, there are 92 cases, and the trend continues to rise.

The areas of application of PDCA are also diverse; although its origin is in engineering, currently, there are many more. The previous Scopus search indicates that the applications in the areas of knowledge have been: Engineering (353), Computer Science (198), Business, Management and Accounting (180), Medicine (180), Social Sciences (97), Decision Sciences (92), Nursing (65), Environmental Science (55), Materials Science (51), Energy (46), Mathematics (46), Chemical Engineering (32), Physics and Astronomy (28), Pharmacology, Toxicology and Pharmaceutics (27), Earth and Planetary Sciences (26), Chemistry (25), among others.

In the same way, it has been found that many authors stand out for making use of PDCA for the improvement of industrial processes and quality, among which stand out with six papers Silva, F. J. G.; with five papers Blagojević, M., Drăgulănescu, N., Ferreira,

Fig. 1.1 PDCA in quality and improvement trends

L. P. and Garza-Reyes, J. A.; with four papers Chină, R., Lontsikh, N. P., Lontsikh, P. A., Micić, Z., Sun, J. and Tsubaki, M., to mention only a few.

Among the journals that publish the most on this topic are Iop Conference Series Materials Science And Engineering (17), Quality Access To Success (12), ACM International Conference Proceeding Series (9), Espacios (9), Advanced Materials Research (8), Advances In Intelligent Systems And Computing (8), Joint Commission Journal On Quality Improvement (8), Applied Mechanics And Mate-rials (7), Journal Of Nursing Care Quality (7), Quality Progress (7), among others.

1.2 Auxiliary Tools

1.2.1 Flow Chart (FC)

A flowchart is a visual tool, an image that shows the workflow for a specific work process. A flowchart uses different symbols to define the type of activity, along with connecting arrows that establish the flow and sequence of a process [14]. For example, it uses boxes or rectangles to represent the activities or steps of the process or task, ovals, or circles to indicate the beginning and end of the process, diamonds to indicate that a decision must be made, and arrows to indicate the sequence of these steps. This facilitates the process's understanding, standardization, and improvement since it makes it possible to identify the process to be analyzed, the total number of activities required, and the beginning and end of the process [15].

The flow diagram offers the following advantages [15, 16]:

1. It allows the identification of the sequence of steps required to perform a task.
2. It allows the relationships between the steps to be identified.
3. It is easy to interpret, follow and remember.
4. It allows to highlight of the transfers, that is to say, the places from which the process flows from one person to another.
5. It allows for detecting problems in the work process under analysis.

1.2.1.1 An Example of Flow Chart

Figure 1.2 shows the flowchart for a manufacturing process of aircraft interior components. Como se puede observar, los componentes deben pasar por nueve estaciones de trabajo, empezando por las estaciones de Pressess y Laminate, para posteriormente pasar por las estaciones de Cutting, Sanding, Bondeo, and Inserts. Si la pieza requiere de pintura, pasa a la estación de Painting; en caso contrario requerirá de papel tapiz, el cual se le coloca en la estación de Decorated. Finamente, todas las piezas terminan en la estación de Assembly.

1.2.1.2 Flow Chart—Brief Literature Review

The words flow chart and flowchart are used interchangeably in the literature, but they always refer to a sequence of activities graphically. In a literature review conducted in Scopus using the following search equation (TITLE-ABS-KEY ("flow chart" OR flowchart) AND TI-TLE-ABS-KEY (quality OR manufacturing)), 2,129 documents have been found that refer to their application in areas associated with quality or manufacturing.

Figure 1.3 illustrates the timeline for the evolution of these concepts (flow chart and quality OR manufacturing). The blue line reports the documents published and the red and dotted line exponential line reports the trend with 96% of fit. These were reported first before the PDCA cycle since the first scientific report appeared in 1968, and since that date, it has been increasing over time. However, no major changes were observed from 1968 to 1982, when scientific production was scarce. From that date until 1990, there is a slight increase until 2000, when the intensity of change is greater, reaching a total of 179 documents in 2021, indicating the great utility found in the industry and other sectors.

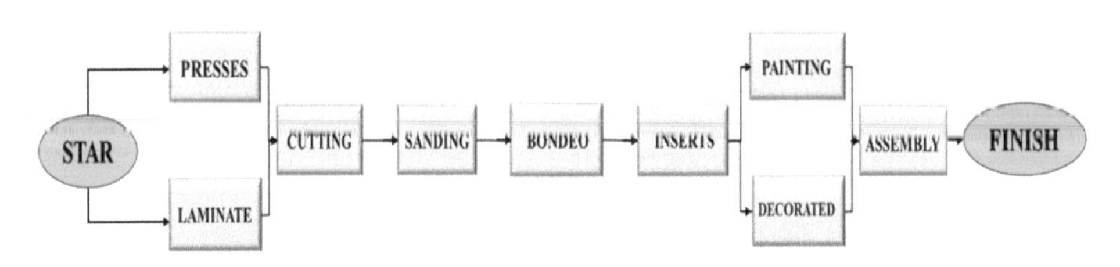

Fig. 1.2 Aircraft interior components manufacturing flowchart

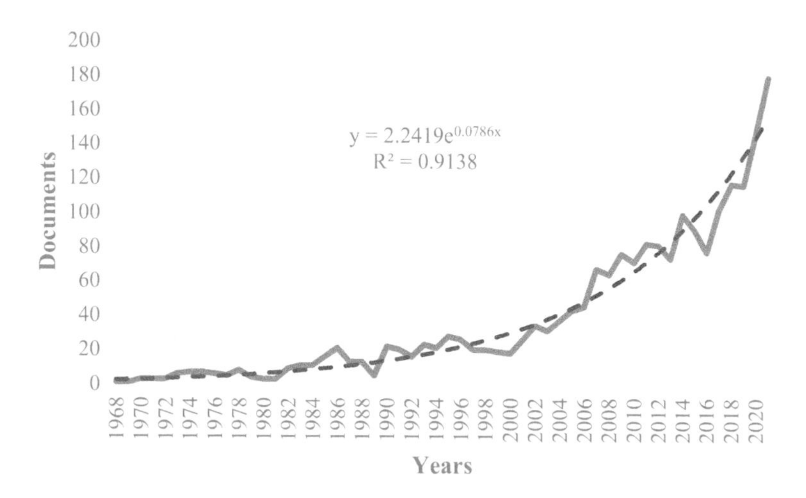

Fig. 1.3 Flow chart in quality and manufacturing trends

The reported papers on flow charts and their combination with quality or manufacturing cover various topics. Although its origins have also been in engineering in general terms, its application recently covers several topics; for example, Engineering (813), Medicine (600), Computer Science (308), Materials Science (143), Environmental Science (142), Business, Management and Accounting (132), Earth and Planetary Sciences (126), Energy (125), Social Sciences (111), Mathematics (108), Physics and Astronomy (105), Biochemistry, Genetics and Molecular Biology (86), Nursing (86), Decision Sciences (76), Agricultural and Biological Sciences (64), Chemical Engineering (59), among others.

Many authors have excelled in this research topic. For example, with 10 documents, there is Nagendrababu, V., with 9 documents, there is Dummer, P. M. H., with 8 documents there is Jayaraman, J., with seven papers there are Anon and Priya, E; with 6 documents, there is Pulikkotil, S. J.; with five documents, there is Formica, M; with four documents, there are Bossuyt, P. M., Felli, L., Kanz, K. G., Kaplan, A. F. H., Moher, D., Murray, P. E., Reitsma, J. B., Sanz-Valero, J., Taype-Rondan, A., Timaná-Ruiz, R., Wan-den-Berghe, C., Waydhas, C. and Zanirato, A., to mention only some of the most important ones.

Since there are many areas of application, the Journals in which it is published are also diverse and among the most important are the following: Advanced Materials Research (18), Proceedings Of SPIE The International Society For Optical Engineering (17), Mining Informational And Analytical Bulletin (16), Gornyi Zhurnal (15), Quality Progress (15), Applied Mechanics And Materials (12), Xibei Gongye Daxue Xuebao Journal Of Northwestern Polytechnical University (11), Iop Conference Series Earth And Environmental Science (10), International Endodontic Journal (19), International Journal Of Advanced Manufacturing Technology (9), Annual Quality Congress Transactions (8), Iop

Conference Series Materials Science And Engineering (8), Plos One (8), Chinese Journal Of Evidence-Based Medicine (7), Dianli Zidonghua Shebei Electric Power Automation Equipment (7), Journal Of Physics Conference Series (7), Medical Physics (7), Advances In Intelligent Systems And Computing (6), among others.

1.2.2 Pareto Diagram

The Pareto diagram emerged when the Italian scientist named Wilfredo Pareto discovered that 20% of the people in Italy received 80% of the wealth [17]. The main function of this diagram is to identify how much some specific factors influence a problem relative to other factors; that is, to identify the best opportunities for improvement [16].

Each bar represents part of a problem or a different category in this diagram. Thus, the Pareto diagram is considered a special type of bar chart [18], in which the frequency of distribution (on the vertical axis) of descriptive data classified into categories (on the horizontal axis) is illustrated [17, 18]. These categories appear in descending order from left to right, while a line represents the cumulative percentage of frequencies. The highest bars represent the categories that contribute the most to the problem. The Pareto diagram has the following advantages [11, 18, 19]:

1. It divides a problem into factors or categories.
2. It prioritizes vital problems over trivial ones by recognizing the key categories contributing most to a specific problem.
3. Guides on where to focus efforts to solve a problem.

Several authors have applied the Pareto diagram in their research. For example, Sharma and Suri [20] apply it to a production process to identify where the most rejected parts occur and give suggestions for improvement to reduce the number of rejects and rework. For their part, Visveshwar et al. [21] provide a systematic and organized implementation of this diagram in a plastic-based production industry to achieve a continuous improvement cycle. In other research, Chokkalingam et al. [22] and Acharya et al. [23] use the Pareto diagram to identify the inclusion of defective sand in foundries, allowing them to determine and check the priority of defects. Finally, Nabiilah et al. [24] apply the Pareto chart in an electroplating process to know the most frequently occurring defects, defining drill bits as the most frequent defect in the process.

1.2.2.1 Example of Pareto Diagram Application

An example of the use of the Pareto diagram is shown below. A company is engaged in manufacturing and delivering products to several customers. However, some customers have returned some of their products. Therefore, it wants to investigate the causes of the returns that have occurred in the last six months. Once the reasons for product returns

Table 1.1 Causes of returns

Causes	Number of occurrences	Cumulative cases	Percentage per unit (%)	Cumulative percentage (%)
Delay in delivery	140	140	28	28
Delay in transport	125	265	25	53
Damaged product	65	330	13	66
Incorrect invoicing	60	390	12	78
Incorrect separation	45	435	9	87
Wrong order	30	465	6	93
Wrong price	20	485	4	97
Other	15	500	3	100
Total	500		100	

are identified, a column is added to the table to keep track of the accumulated cases, another column where the unit percentage of each case is measured, and the accumulated percentage column. Table 1.1 shows the causes of returns, the number of occurrences for each reason, the accumulated cases, and their corresponding percentages.

The Pareto chart can be constructed with the data in Table 1.1, which is presented in Fig. 1.4. As can be seen, to reduce the problem of product returns, the company must create an action program to reduce delivery delays from the factory and the carrier. With this strategy, 53% of the problem will be solved.

1.2.2.2 Pareto Diagram—Brief Literature Review

Since Pareto diagrams are widely used in problem-solving as auxiliary tools, a search has been performed in Scopus with the following search equation ALL ("Pareto diagram"). In this case, there is no emphasis on a specific area since its application is in several sectors, and it is a generic tool that is easily adapted and understood. The search results indicate 449 documents in which the tool is mentioned.

Figure 1.5 shows the timeline in which documents mentioning and using this tool have been published in the blue line, to which a third-order trend line has been fitted, dotted and red, with a 94.47% fit. In general terms, it is observed that the first time that the appearance of the Pareto diagram in a scientific document was documented in Scopus was in 1983, and for more than two decades, its use remained very low. It is in 2007 when the curve begins an exponential growth with ups and downs, but by 2021 there are

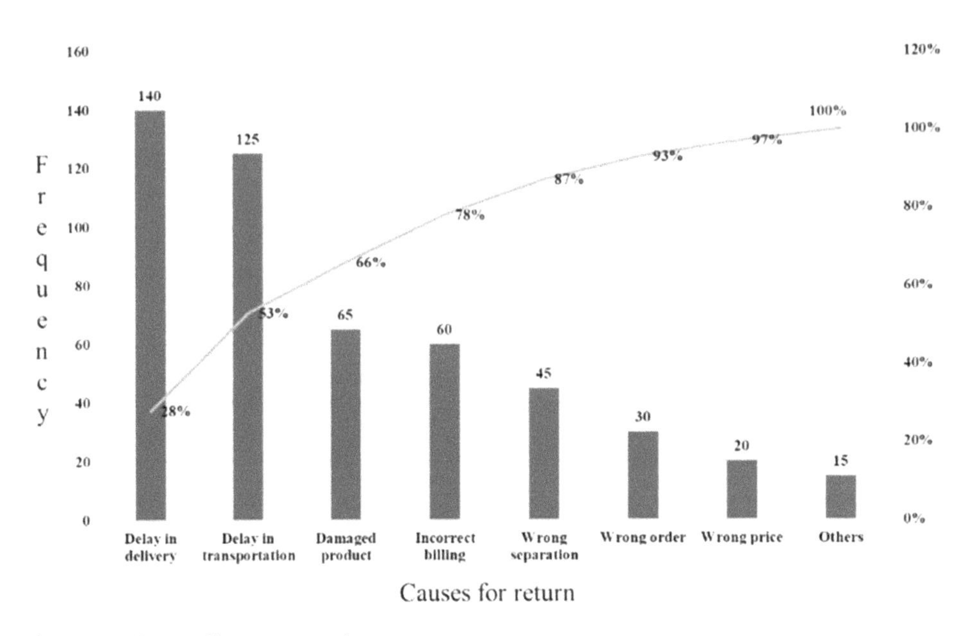

Fig. 1.4 A Pareto diagram example

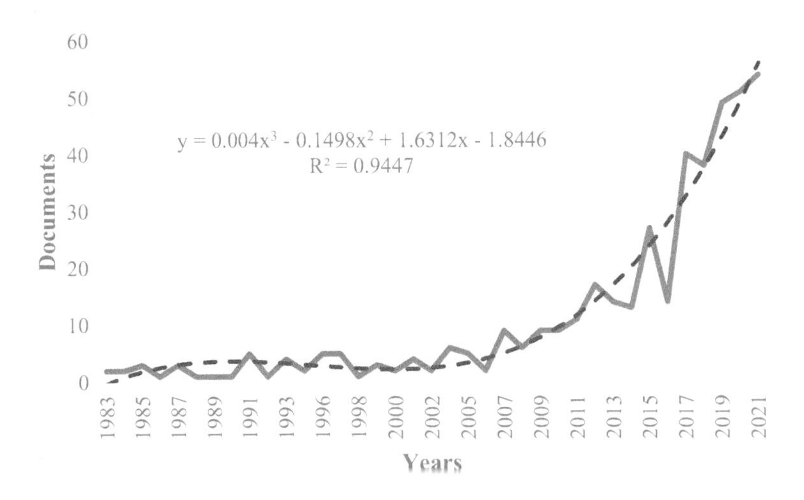

Fig. 1.5 Pareto diagram usage timeline

already 54 documents that refer to this tool. In addition, the adjusted trend line indicates that it will continue to be used, mainly due to its simplicity and ease of interpretation.

The use of Pareto diagrams has been reported in different areas of research to support problem-solving. In the Scopus search, the following categories and values have been found: Engineering (238), Business, Management and Accounting (61), Computer Science

(60), Materials Sci-ence (55), Agricultural and Biological Sciences (50), Environmental Science (46), Medicine (44), Decision Sciences (36), Social Sciences (34), Physics and Astronomy (30), Energy (29), Mathematics (27), Chemical Engineering (25), Earth and Planetary Sciences (19), Chemistry (16), Biochemistry, Genetics and Molecular Biology (14), Pharmacology, Toxicology and Pharmaceutics (9), Immunology and Microbiology (7), Nursing (7), among others.

Many authors have also used the Pa-challenge diagram in different areas, contributing different documents and approaches. For example, with seven papers, there are Cordeiro, T. B. C., Morga-do, C. and Varzakas, T; with six papers, there are Arvanitoyannis, I. S. and Varzakas, T. H.; with four papers there are Certa, A., Djekic, I., La Fa-ta, C. M., Ozilgen, S., Psomas, E., Purba, H. H., Suwanasri, C., Suwanasri, T., An-tony, J., Dumitrascu, A. E., Enea, M., Galante, G. M., Giannou, V., Jaqin, C., Pecherskaya, E. A., Saroso, D. S., Tomasevic, I., Torres-Palma, R. A. and Tzia, C. However, it is important to mention that there are many more authors with two or one paper.

The main journals in which the use of this tool is reported are very varied in focus, among which are Iop Conference Series Mate-rials Science And Engineering (20), Iop Conference Series Earth And Environmental Science (9), Aip Conference Proceedings (7), Advanced Materials Research and Journal Of Physics Conference Series (5), Applied Mechanics And Materials, Critical Reviews In Food Science And Nutrition, International Journal Of Food Science And Technology, International Journal Of Productivity And Quality Management and Sustainable Energy Technologies And Assessments (4), Acta Technologica Agriculturae, Annual Quality Congress Transactions, Applied Thermal Engineering, Food Control, IEEE Access, International Journal Of Scientific And Technology Research, Journal Europeen Des Systemes Automatises, Pharmaceutical Care And Research and TQM Journal (3).

1.2.3 Process Flow Chart (PFC)

The process flow diagram helps to detect non-productive hidden costs in the form of waste, such as distances traveled, delays and temporary material stockpiles to which an item is exposed as it travels through the plant. Once these wastes are identified, corrective actions can be taken to minimize them and reduce their costs [25, 26].

Process flow diagrams illustrate a product's operations, inspections, transports, delays, and storage. Various symbols represent these; for example, a circle signifies an operation, a square represents an inspection, and a small arrow signifies transportation, which is defined as the action of moving an object from one place to another, except when the movement takes place during the normal course of an operation or inspection. A capital letter D stands for a delay, which occurs when a part cannot be processed immediately at the next workstation, and finally, an equilateral triangle stands for storage, which occurs when a part is stored and protected in a certain location [25, 26].

The process flow diagram provides the details of events involving a product or material and is identified by a title and additional accompanying information (part number, diagram number, process description, current or proposed method, date, and name of the person who produced the diagram). This tool facilitates the elimination or reduction of hidden costs of a component, and since it clearly shows all transports, delays and storage, the information it provides can result in a reduction in the quantity and duration of these items [25, 26].

1.2.3.1 Example of PFC Application

Table 1.2 shows an example of a process flow chart in which the different tasks carried out during the assembly of a machine in a carton company can be observed: operation, transport, storage, inspection and delay [27]. As can be seen, the analyzed activity is changing manufacturing order (MO) to reduce the setup time of the analyzed machine. In addition, it can also be observed that the process is composed of 23 tasks, of which 12 are operations, and 11 are transports. From this example, it can be quickly observed that transports tasks that do not add value; therefore, a project should be carried out to eliminate them.

1.2.3.2 PFC—Brief Literature Review

The PFC is sometimes also called process flowchart or process flow diagram, so a search was performed in Scopus with the following search equation ALL ("process flowchart" OR "process flow chart" OR "process flow diagram"), obtaining 1543 documents in which mention is made of this tool, which indicates its wide use in the identification of the beginning and end of processes.

Figure 1.6 shows the timeline of the number of documents and possible names according to the search equation found in scientific papers. The blue line indicates the number of documents that refer to the tool, and the blue line indicates a third-grade line with a fit of 96.86%, which indicates that its use has an increasing trend.

It can be seen that this type of graph appeared for the first time in 1965 and was little used until 1979 when an increase was observed until 1983. After that, there was a slight decrease until 1992, when it began to increase again, totaling 117 documents in 2021.

The use of PFCs, although also originating in industry, is currently universally applied when it is desired to represent the beginning and end of all the activities of a process and the search in Scopus indicates that the main areas are Engineering (678), Chemical Engineering (292), Energy (253), Environmental Science (214), Computer Science (204), Earth and Planetary Sciences (193), Materials Science (165), Chemistry (165), Chemistry (165), Materials Science (165), Earth and Planetary Sciences (193), Materials Science (165), Chemistry (160), Business, Management and Accounting (111), Physics and Astronomy (107), Mathematics (94), Medicine (78), Social Sciences (55), Decision

Table 1.2 Example of process flow chart for a machine assembly task. Adapted from Roriz et al. [27]

Process flow chart		Summary				
Chart 1	Sheet 1 of 1	Activity	Values			
Objective: To reduce the setup time of the machine under study Activity: Change manufacturing order Location: Countertop section Operator: 1 Equipment: Scaffolding 1	Operations	⬭	12			
	Transportation	⇨	11			
	Storage	▽	0			
	Inspection	□	0			
	Waiting	D	0			
	Symbols					
No	Description of operation	⬭	⇨	▽	□	D
1	Stop the machine to change MO	⬭				
2	Arrange the plans that are left over from the previous MO		⇨			
3	Record the quantity produced in the previous MO	⬭				
4	Check which is the next MO	⬭				
5	Cut the micro in the machine inlet	⬭				
6	Remove the micro and place it on the waste pallet		⇨			
7	Wrap the remaining bobbin and identify it	⬭				
8	Lower the shaft coil	⬭				

(continued)

Table 1.2 (continued)

Process flow chart		Summary				
9	Pack the micro coil in the hallway		⇨			
10	Search for a pallet truck		⇨			
11	Find the plans and transport them to the machine		⇨			
12	Arrange the pallet truck		⇨			
13	Load the machine with the planes to be	○				
14	Find the bobbin and transport it to the machine		⇨			
15	Raise reel to shaft	○				
16	Remove the 1st turn from the reel	○				
17	Take the 1st turn to the waste pallet		⇨			
18	Run the micro through the machine inlet	○				
19	Tuning machine (wheels and suction cups)	○				
20	Pass the micro through the machine rollers	○				
21	Search for a pallet truck		⇨			
22	Arrange the pallet with the contracted planes		⇨			
23	Turn the machine on and remove the trash		⇨			
Total	12	11	0	0	0	

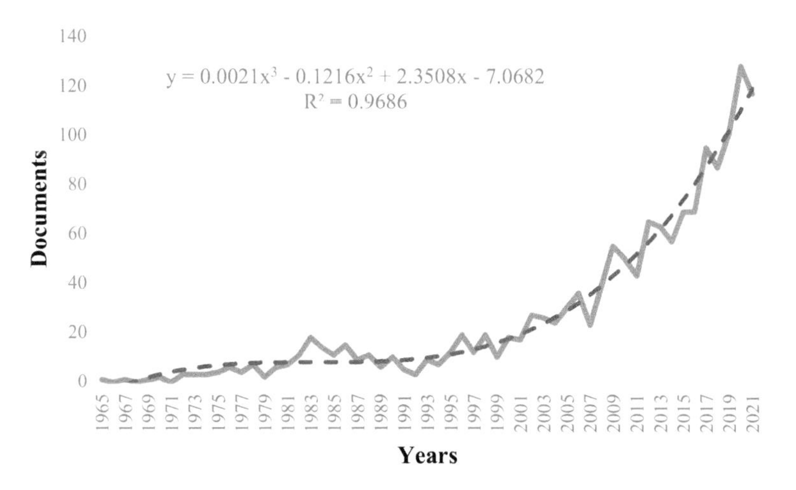

Fig. 1.6 PFC usage timeline

Sciences (48), Biochemistry, Genetics and Molecular Biology (47), Agricultural and Biological Sciences (43), Economics, Econometrics and Finance (30), Nursing (11), Health Professions (10), among others.

In the same way, many authors report using CBPs as an auxiliary to represent process that indicates a sequence of activities. Some of the most important authors are Poch, M. (12), Garrido-Baserba, M. (11); Anon, Gao, Z. K. and Jin, N. D. (7); Sagdatullin, A. (6); Castillo, A., Comas, J. and Maldonado, P. (5); Chen, L. J., Choi, J. O., Cozzani, V., Ferrão, L., Geller, V. E., Godini, H. R., Landucci, G., Lee, B., Lim, H., Lu, H. B., Molinos-Senante, M., Nakagaki, T., Reif, R., Tugnoli, A., Wang, S. L., Wozny, G., Yang, F., Yang, S. R. and Zhou, W. X. (4); Adi, V. S. K., Babazadeh, F., Babokin, G. I., Barbe, M. C., Birk, W., Boland, G. W. L., Chang, C. T., Correia, A. R., Daneault, C., Daoutidis, P., Dong, J., Ermida, P., Gazaleeva, G. I., Ghezel-Ayagh, H., Glebov, A. V., Haddadi, A. M., Hassim, M. H., Hayashi, H., Holley, C. A. (3); among others.

Some of the journals that publish the most about CBP include Gornyi Zhurnal (38), Mining Informational And Analytical Bulletin (30), Iop Conference Series Earth And Environmental Science (19), Chemical And Petroleum Engineering (14), Chemical Engineering Progress and Industrial And Engineering Chemistry Research (13), Journal Of Cleaner Production and Theoretical Foundations Of Chemical Engineering (12), Journal Of Mining Science and Tsvetnye Metally (11), Chemical Engineering Transactions (9), Iop Conference Series Materials Science And Engineering, Journal Of Physics Conference Series and Lecture Notes In Computer Science Including Subseries Lecture Notes In Artificial Intelligence And Lecture Notes In Bioinformatics (8), Chemical Engineering Research And Design, Computers And Chemical Engineering, Energies, Journal Of Natural Gas Science And Engineering, SAE Technical Papers, Thermal Engineering (7), Aiche Journal, Aip Conference Proceedings, Computer Aided Chemical Engineering,

Energy, Metallurgist, Advances In Intelligent Systems And Computing, Chemical Processing, China Petroleum Processing And Petrochemical Technology, Environmental Science And Technology, IFIP Advances In Information And Communication Technology, Jisuanji Jicheng Zhizao Xitong Computer Integrated Manufacturing Systems CIMS, Proceedings Of The ASME Turbo Expo, Ugol (5), entre otros.

1.2.4 Ishikawa Diagram

The Ishikawa diagram, also known as the fishbone diagram, or cause and effect diagram, was developed by Kaoru Ishikawa, a Japanese quality control leader [28]. Such a diagram is a graphical tool used to concisely identify the causes and effects of a problem. Because the causes of the problem are prioritized, it is possible to identify the sources of the problem. In addition, it can also be used as an analytical tool in project management and quality control [29].

Typically, this diagram results from a brainstorming session in which problem solvers make suggestions. The main goal is represented by the head of a fish, while the main factors are represented by spines attached to the backbone. Subsequently, secondary factors are added as stems, and so on. Creating Ishikawa diagrams generally stimulates discussion and promotes understanding of a complex problem [28].

Typically, the main causes are subdivided into five or six major categories—human, machine, method, material, environmental, and measurement- each further subdivided into sub-causes. The process continues until all possible causes are identified and listed. A good diagram will have several levels of spines and provide a good picture of the problem and the factors contributing to its existence [25].

1.2.4.1 Ishikawa Diagram Example

Figure 1.7 shows an example of an Ishikawa diagram for a customer complaint problem in a manufacturing company, in which the diagram's structure can be clearly seen. In this case, six main causes are observed, associated with the workforce, machinery, measurement systems, method, material, and environment, which have other sub-causes or ramifications. In this case, since it is a service, the material and the work environment are not analyzed since they are unimportant.

1.2.4.2 Ishikawa Diagram—Brief Literature Review

The Ishikawa diagram is also known as the fishbone diagram or cause and effect diagram, so a search was performed in Scopus to co-know the uses and mentions it has received and to know its scientific impact in problem-solving. The search equation was ALL ("Ishikawa diagram" OR "Fishbone Diagram" OR "Cause and Effect Diagram" OR "Cause and Effect Diagram"), and a total of 2445 documents were found.

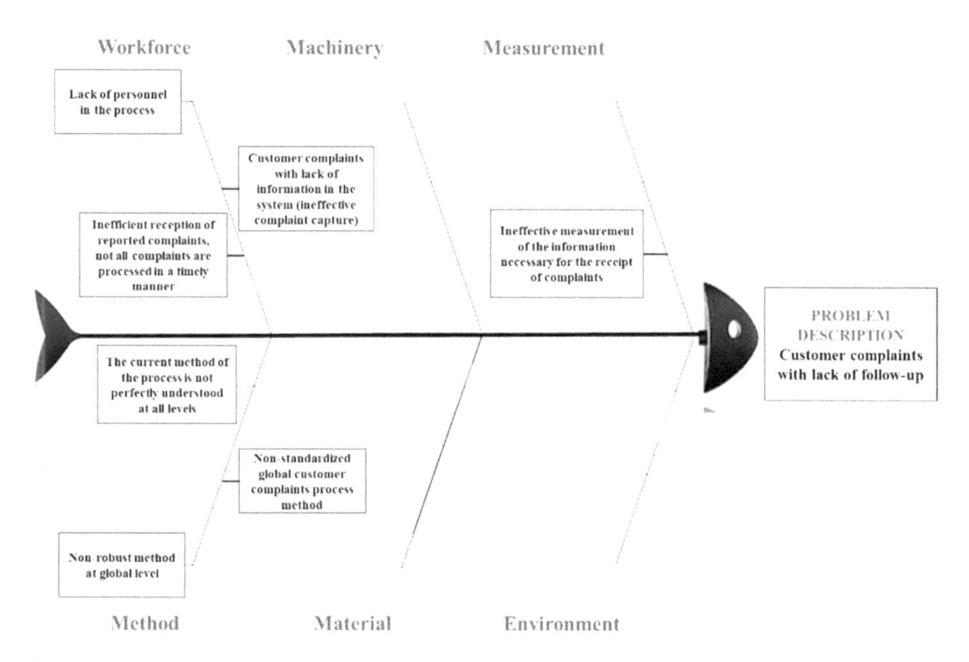

Fig. 1.7 Ishikawa diagram for a customer complaint problem

Figure 1.8 indicates the timeline of the use of the Ishikawa diagram in scientifically reported problem-solving. It can be seen that the first time the tool was mentioned was in 1969, and its use remained almost null until 1984, when two mentions appeared. From that date on, there was a gradual but slow increase until 2004, when there were 17 mentions. It is from 1985 when an exponential increase in its use is observed, with 357 mentions in the year 2021, which indicates that it is a vital tool, easy to use and manage in the solution of problems.

The applications of the fish diagram have been in different areas of science, among which are Engineering (1131), Business, Management and Accounting (499), Computer Science (430), Medicine (341), Decision Sciences (250), Materials Science (240), Social Sciences (226), Environmental Science (187), Mathematics (139), Physics and Astronomy (134), Energy (119), Chemical Engineering (116), Chemistry (102), Agricultural and Biological Sciences (95), Pharmacology, Toxicology and Pharmaceutics (91), Earth and Planetary Sciences (87), Nursing (78), Biochemistry, Genetics and Molecular Biology (77), Economics, Econometrics and Finance (63), Health Professions (36)), Arts and Humanities (23), Multidisciplinary (19), Psychology (16), among others.

The authors who have used the Ishikawa diagram to solve problems and report them in their scientific reports vary. The following are the authors and papers, such as Noorzai, E. (23), Pacana, A. (20), Meinrath, G. (16), Kenett, R. S. (13), Lis, S. and Siwiec, D. (12), Coccia, M. and Rios, N. (11), Arvanitoyannis, I. S. and Varzakas, T. H. (10), Kumar, S.,

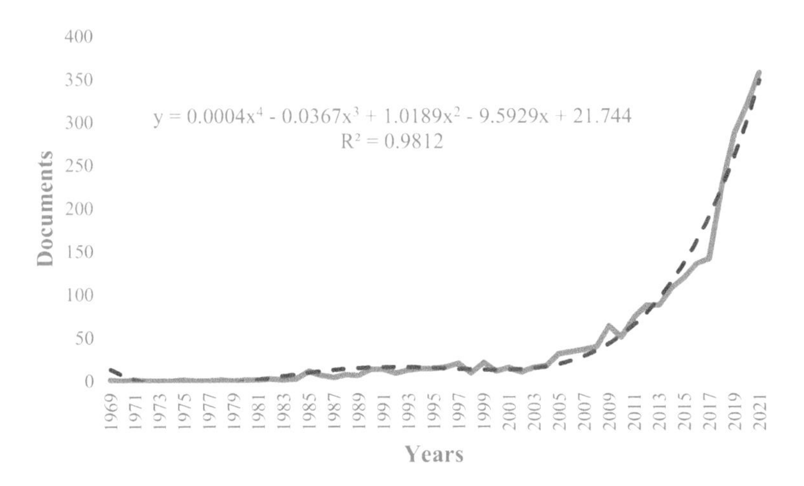

Fig. 1.8 Ishikawa diagram timeline

Seaman, C., Varzakas, T. and Zhao, D. (9), Kalinowski, M. and Meyer, V. R. (8), Antony, J., Chen, J., Cordeiro, T. B. C., Czerwińska, K., Dong, H., Freire, S., Gijo, E. V., Liu, Y., Morgado, C., Wang, W. and Wang, Y. (7), among others.

These authors have published their papers in several journals, conferences or book series and the most important ones and their numbers are listed below; for example, Iop Conference Series Materials Science And Engineering (61), Iop Conference Series Earth And Environmental Science (34), Quality Progress (25), Advances In Intelligent Systems And Computing (24), Journal Of Physics Conference Series (21), Aip Conference Proceedings (19), International Journal Of Productivity And Quality Management (18), ACM International Conference Proceeding Series (15), Advanced Materials Research (15), Annual Quality Congress Transactions (14), Lecture Notes In Computer Science Including Subseries Lecture Notes In Artificial Intelligence And Lecture Notes In Bioinformatics (14), Journal Of Cleaner Production, Lecture Notes In Mechanical Engineering, SAE Technical Papers and Sustainability Switzerland (13), Quality Access To Success (12), among others.

1.2.5 5's

The 5's are five Japanese principles whose names begin with S and aim to achieve a clean and tidy workplace. They represent a program that consists of developing activities of order/cleanliness and detection of anomalies in workplaces. They offer the advantages that, due to their simplicity, their application allows the participation of workers at all levels and that they truly coexist, improving the work environment, people's safety, and productivity [30].

The pillars of the 5S program are (1) classify (seiri), (2) tidy up (seiton), (3) polish (seiso), (4) standardize (seiketsu), and (5) conserve (shitsuke) [25, 30]. Sorting focuses on removing all unnecessary items from the workplace to leave only what is fundamental and necessary. Tidying up involves arranging necessary items in such a way that they are easy to find and use. Once the clutter is removed, polishing ensures cleanliness and neatness. Once the first three pillars of the 5's have been implemented, standardization maintains order and consistency of housekeeping tasks and methods. Finally, the con-serving step maintains the entire process on an ongoing basis [25, 31].

1.2.5.1 5's—Example of Its Application

Figure 1.9 shows an example of the application of the 5's in a plastic products manufacturing plant [32]. As can be seen, the upper part (enclosed in red) shows the situation of disorder that existed in different areas before implementing the 5's tool, while the lower part shows the ordered situation, where each material, component or hand tool is in the place that was assigned to it after implementing the 5's.

1.2.5.2 5's—Brief Literature Review

The reports of the application of the 5's are very diverse and not only in the industry, where it has its origin but currently comprises several areas of knowledge and has proven to be a powerful tool that supports standardization and industrial improvement. A search was performed in Scopus with the search equation (ALL (5's) AND ALL (seiri)), which has been integrated with the word SEIRI to ensure that it refers to the 5s applied in industry and services since it is also an acronym to define other words. A total of 1222 documents have been found that refer to the 5s.

Figure 1.10 illustrates the timeline for the 5s and shows that the first time these two words appeared together was in 1948, but the word was not used again in any report for

Fig. 1.9 Example of application of the 5's in a plastic products manufacturing plant [32]

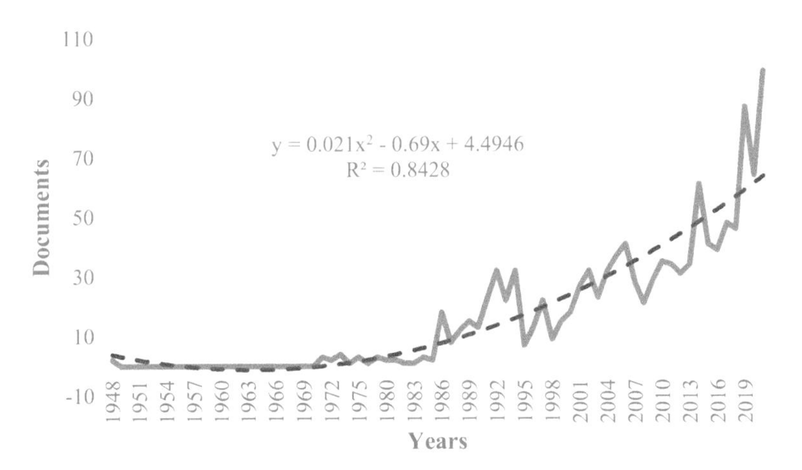

Fig. 1.10 5s applications—timeline

more than two decades. It was not until 1971 that three documents appeared again, and for a decade, an average of two or three documents were generated per year, and it was not until 1986 that the tool began to grow exponentially, reaching 99 documents in the year 2021.

The areas of application of the 5s are very diverse, and although its application began in the industry, it is observed that other areas have seen its usefulness and are now leaders. The following are the areas and the number of documents in which they appear, such as Medicine (605), Biochemistry, Genetics and Molecular Biology (325), Agricultural and Biological Sciences (147), Health Professions (147), Social Sciences (144), Engineering (96), Arts and Humanities (74), Environmental Science (68), Nursing (45), Pharmacology, Toxicology and Pharmaceutics (45), Chemistry (43), Materials Science (42), Business, Management and Accounting (39), Neuroscience (36), Immunology and Microbiology (34), Computer Science (32), Physics and Astronomy (29), Chemical Engineering (26), Mathematics (19), Economics, Econometrics and Finance (18), Earth and Planetary Sciences (17), among others.

The main authors reporting the use of this tool for the standardization of processes or set of activities and workplaces are very varied, and the following are the most important and the number of documents, such as Shibata, K. (50); Fukui, T. (22); Jones, A. M. (20); Onodera, M. (17); Matsuo, H. (14); Hanberry, B. B., Hughson, R. L. and Tanaka, K. (10); Barstow, T. J., Hayakawa, T., Levine, B. D. and Watanabe, T. (9); He, H. S., Nour-bakhsh, M., Poole, D. C. and Sakamoto, K. (8); Katsuura, T., Koga, S., Rossiter, H. B., Scheuermann, B. W. and Seiri, P. (7); Abdolvahabi, Z., Ebara, S., Faigenbaum, A. D., Hesari, Z., Kondo, N., Perrey, S., Sairyo, K., Tsuge, H., Vanhatalo, A., Xu, X., Bailey (5); among others.

The journals that have published the most on these topics are also very diverse and serve a variety of disciplines. Listed below are the top journals and the number of papers they have published, such as Agricultural And Bio-logical Chemistry (45), Journal Of Applied Physiology (35), European Journal Of Applied Physiology (26), Medicine And Science In Sports And Exercise (21), In-ternational Journal Of Sports Medicine (15), Journal Of Strength And Condition-ing Research (14), Bioscience Biotechnology And Biochemistry (12), Journal Of Cellular Biochemistry (11), Journal Of Sports Medicine And Physical Fitness (9), Journal Of Thermal Biology (9), Sports Medicine (9), Journal Of Physiological An-thropology (8), European Journal Of Applied Physiology And Occupational Physiology (7), Japanese Journal Of Physical Fitness And Sports Medicine (7), Journal Of Cellular Physiology (7), Journal Of Nutritional Science And Vitaminology (7), Journal Of Sports Sciences (7), American Journal Of Physiology Regulatory Inte-grative And Comparative Physiology (6), Biological And Pharmaceutical Bulletin (6), Building And Environment (6), Frontiers In Physiology (6), Journal Of Physio-logical Anthropology And Applied Human Science (6), Journal Of Physiology (6), Journal Of The Japan Research Association For Textile End Uses (6), Lighting Re-search And Technology (6), among others.

1.3 Conclusions

The PDCA cycle and the different auxiliary tools reported in this chapter are excellent resources for analyzing and treating information and monitoring it in manufacturing process improvement projects. In the case of the 5's, this is an essential tool for the good performance of a worker, not only in manufacturing processes but in any work sector. On the other hand, the Ishikawa diagram can be seen as the starting point for an improvement project since it allows for detecting problems and their causes, which gives rise to the project. Concerning the Pareto diagram, it increases the possibility of making correct decisions from the beginning of the project by highlighting the main causes of the problem. The flow chart and flow process chart detect improvement opportunities (waste) in a production process. Finally, the PDCA cycle is an excellent method for managing and controlling improvement processes, which is why it is widely used worldwide.

References

1. Strotmann C, Göbel C, Friedrich S, Kreyenschmidt J, Ritter G, Teitscheid P et al (2017) A participatory approach to minimizing food waste in the food industry—a manual for managers. Sustainability [Internet] [cited 2018 Oct 10] 9(1):66. http://www.mdpi.com/2071-1050/9/1/66
2. de Souza JM (2016) PDCA and lean manufacturing: case study in appliance of quality process in alpha graphics (in Portuguese). J Leg Bus Sci 17:11–17

3. Silva AS, Medeiros CF, Vieira RK (2017) Cleaner production and PDCA cycle: practical application for reducing the Cans loss index in a beverage company. J Clean Prod [Internet] [cited 2018 Oct 10] 150:324–38. https://www.sciencedirect.com/science/article/pii/S0959652617304687
4. Isniah S, Purba HH, Debora F (2020) Plan do check action (PDCA) method: literature review and research issues. J Sist dan Manaj Ind [Internet] [cited 2022 Feb 15] 4(1):72–81. https://e-jurnal.lppmunsera.org/index.php/JSMI/article/view/2186
5. Sangpikul A (2017) Implementing academic service learning and the PDCA cycle in a marketing course: contributions to three beneficiaries. J Hosp Leis Sport Tour Educ [Internet] [cited 2018 Oct 5] 21:83–87. https://www.sciencedirect.com/science/article/pii/S1473837617300412
6. Kholif AM, Abou El Hassan DS, Khorshid MA, Elsherpieny EA, Olafadehan OA (2018) Implementation of model for improvement (PDCA-cycle) in dairy laboratories. J Food Saf [Internet] [cited 2018 Oct 29] 38(3):e12451. http://doi.wiley.com/10.1111/jfs.12451
7. Schneider PD (1997) FOCUS-PDCA ensures continuous quality improvement in the outpatient setting. In: Oncology nursing forum, p 966
8. Muhammad S (2015) Quality improvement of fan manufacturing industry by using basic seven tools of quality: a case study. Int J Eng Res Appl 5(4):30–35
9. Maruta R (2012) Maximizing knowledge work productivity: a time constrained and activity visualized PDCA cycle. Knowl Process Manag [Internet] [cited 2018 Oct 10] 19(4):203–214. https://onlinelibrary.wiley.com/doi/abs/10.1002/kpm.1396
10. Gorenflo G, Moran JW (2009) The ABCs of PDCA [Internet]. Accreditation Coalition, Minnesota, p 7. http://www.phf.org/resourcestools/Documents/ABCs_of_PDCA.pdf
11. Realyvásquez-Vargas A, Arredondo-Soto KC, Carrillo-Gutiérrez T, Ravelo G (2018) Applying the plan-do-check-act (PDCA) cycle to reduce the defects in the manufacturing industry. A case study. Appl Sci 8(11):1–17
12. Mutafelija B, Stromberg H (2003) Systematic process improvement using ISO 9001:2000 and CMMI [Internet], 1st ed. Norwood, Massachusetts: Artech House; [cited 2022 Mar 22], pp 1–300. https://books.google.com.mx/books?id=F1oq7FCp-MgC&pg=PA16&dq=pdca+cycle&hl=es&sa=X&ved=2ahUKEwjire3cvtv2AhVGJEQIHd_fALQQ6AF6BAgLEAI#v=onepage&q=pdcacycle&f=false
13. Khanna HK, Laroiya SC, Sharma DD (2010) Quality management in Indian manufacturing organizations: some observations and results from a pilot survey. Brazilian J Oper Prod Manag [Internet] [cited 2018 Oct 10] 7(1):141–162. https://bjopm.emnuvens.com.br/bjopm/article/view/V7N2A7
14. Vargas-Guarategúa J (2020) Diagrama De Flujo/flowchart [Internet]. Diagrama de flujo/Flowchart. [cited 2022 Mar 23]. https://es.linkedin.com/pulse/diagrama-de-flujo-flowchart-javier-vargas-guarategúa#:~:text=Seusan-ampliamente-en-numerosos, décadas de 1920 y 1930.
15. Joiner Associates I (1995) Flowcharts: plain and simple [Internet]. Reynard S (ed). Oriel Incorporated, Madison, 118 pp. https://books.google.com.mx/books?id=q0dDbdUuGJoC&printsec=frontcover&hl=es&source=gbs_ge_summary_r&cad=0#v=onepage&q&f=false
16. Webber L, Wallace M (2011) Quality control for dummies [Internet]. Indianapolis, Wiley, 384 pp. https://books.google.com.mx/books?id=9BWkxto2fcEC&printsec=frontcover&dq=Weber+and+Wallace.+(2011).+Quality+Control+for+Dummies&hl=es&sa=X&ved=0ahUKEwidsJWE2vzdAhWHhFQKHXoUBwUQ6AEIKjAA#v=onepage&q=Weber+and+Wallace.Quality Control for Dummies&f=f
17. Hosein BM, Hajizadeh R, Dehghan SF, Aghababaei R, Jafari SM, Koohpaei A (2018) Investigation of the accidents recorded at an emergency management center using the pareto chart:

a cross-sectional study in Gonabad, Iran, During 2014–2016. Heal Emergencies Disasters Q [Internet]. [cited 2018 Oct 10] 3(3):143–150. https://doi.org/10.29252/NRIP.HDQ.3.3.143
18. Joiner Associates I (1995) Pareto charts: plain & simple [Internet]. In: Reynard S (ed) Oriel Incorporated, Madison, 130 pp. https://books.google.com.mx/books?id=Mubz8xTERqEC&printsec=frontcover&dq=Pareto+Charts:+Plain+%26+Simple&hl=es&sa=X&ved=0ahUKEwjU3tmZ2fzdAhWNLHwKHRxvDw8Q6AEIJzAA#v=onepage&q=ParetoCharts%3APlain%26Simple&f=false
19. Joiner Associates I (1995) Introduction to the tools: plain & simple: learning and application guide. In: Reynard S (ed) Oriel Incorporated, Madison, 84 pp
20. Sharma H, Suri NM (2017) Implementation of quality control tools and techniques in manufacturing industry for process improvement. Int Res J Eng Technol [Internet] [cited 2018 Oct 30] 4(5):1581–1587. www.irjet.net
21. Visveshwar N, Vishal V, Vimal-Samsingh R, Pragadish-Karthik (2017) Application of quality tools in a plastic based production industry to achieve the continuous improvement cycle. Qual Access Success 18(157):61–64
22. Chokkalingam B, Raja V, Anburaj J, Immanual R, Dhineshkumar M (2017) Investigation of shrinkage defect in castings by quantitative Ishikawa diagram. Arch Foundry Eng [Internet] [cited 2018 Oct 30] 17(1):174–178. https://content.sciendo.com/view/journals/afe/17/1/article-p174.xml
23. Acharya SG, Sheladiya MV, Acharya GD (2018) An application of PARETO chart for investigation of defects in FNB casting process. J Exp Appl Mech [Internet]. [cited 2018 Oct 30] 9(1):33–39. http://engineeringjournals.stmjournals.in/index.php/JoEAM/article/view/392
24. Nabiilah AR, Hamedon Z, Faiz MT (2018) Improving quality of light commercial vehicle using PDCA approach. J Adv Manuf Technol [Internet] [cited 2018 Oct 29] 12(1(2)):525–534. http://journal.utem.edu.my/index.php/jamt/article/view/4310
25. Freivalds A, Niebel BW (2014) Niebel's methods, standards, and work design, 13th edn. McGraw-Hill, New York, p 735
26. Niebel BW, Freivalds A (2009) Ingeniería industrial. Métodos, estándares y diseño del trabajo [Internet], 12th ed. McGraw-Hill Interamericana de España, Mexico City, 744 pp. https://books.google.com.mx/books?id=GI-dQwAACAAJ&dq=Ingeniería+industrial.+Métodos,+estándares+y+diseño+del+trabajo&hl=es&sa=X&redir_esc=y
27. Roriz C, Nunes E, Sousa S (2017) Application of lean production principles and tools for quality improvement of production processes in a carton company. Procedia Manuf 1(11):1069–1076
28. Office of Government Commerce GB (2007) Service operation [Internet]. The Stationery Office, London, 263 pp. https://books.google.com.mx/books?id=ZayAKN3hyUoC&dq=ishikawa+diagram&hl=es&source=gbs_navlinks_s
29. de Saeger A, Feys B (2015) The Ishikawa diagram for risk management: anticipate and solve problems within your business [Internet]. 50 Minutes, 32 pp. https://books.google.com.mx/books?id=0fuOCgAAQBAJ&printsec=frontcover&dq=The+Ishikawa+Diagram+for+Risk+Management:+Anticipate+and+solve+problems+within+your+business&hl=es&sa=X&ved=0ahUKEwjs2enShqDeAhUowlQKHRDGA4YQ6AEILDAA#v=onepage&q=The+Ishikawa+Diagram
30. Sacristán FR (2005) Las 5S. Orden y limpieza en el puesto de trabajo [Internet]. FC Editorial, Madrid, Spain, 167 pp. https://books.google.com.mx/books?id=NJtWepnesqAC&dq=5s&hl=es&source=gbs_navlinks_s

31. Aldavert J, Vidal E, Lorente JJ, Aldavert X (2018) 5S para la mejora continua: La base del Lean [Internet], 3rd ed. Alda Talent, Spain, 234 pp. https://books.google.com.mx/books?id=KEzcDwAAQBAJ&dq=5s&hl=es&source=gbs_navlinks_s
32. Ribeiro P, Sá JC, Ferreira LP, Silva FJG, Pereira MT, Santos G (2019) The impact of the application of lean tools for improvement of process in a plastic company: a case study. Procedia Manuf 1(38):765–775

2 Case Study 1. Reducing Industrial Waste Disposal Costs

2.1 Introduction

Currently, a large amount of waste is generated every day, such as electronic/electrical items, manufacturing scrap, discarded building materials, polymers from daily needs, while their treatment is slow [1]. Due to this, the need for high-performance materials has become latent in high-tech sectors, such as the manufacturing industry [2]. Improving the efficiency of materials helps to reduce the volume of industrial waste, as well as the consumption of resources [2] and efficiency in the use of material resources (waste, garbage) helps to improve productivity, reduce costs, generate economic opportunities, increase efficiency and effectiveness, as well as the competitiveness of the company. For this reason, material resource efficiency is considered one of the most important strategies to generate value in the industry [3]. However, despite the advantages offered by this strategy, even today, several companies around the world continue to make poor use of resources (waste, garbage), as recent research has shown.

For example, Jain and Gupta [4] mention that developed nations are working hard to deal with the problem related to textile waste, as is the case in India, a country in which garment industries have been generating textile waste for about 40 years, and also do not know strategies to generate an effective and long-lasting reverse flow to reuse waste materials, or what is known today as reverse logistics.

On the other hand, Mativenga et al. [5] mention that there is currently no success worldwide in recycling waste materials. These authors conducted a study on waste management in 22 companies in South Africa and indicated that 53% of these companies threw their waste in a landfill, which generated costs for the companies. Similarly, research on the recovery of high-performance fibers has increased worldwide, particularly carbon

A. Realyvásquez Vargas et al., *The PDCA Cycle for Industrial Improvement*, Synthesis Lectures on Engineering, Science, and Technology,
https://doi.org/10.1007/978-3-031-26805-2_2

fiber, due to the tremendous increase in its use and projected demand in various industries, such as automotive (high volume; high production rate; medium service life) and aerospace (high volume, low production rate, long service life) [6].

Nilakantan and Nutt [7] note that a large amount of thermoset and uncured substance-impregnated waste is generated in manufacturing processes, including waste from layer cuttings and off-spec material. However, although techniques for recycling cured composite wastes and recovering carbon fiber are well established and commercialized, few efforts are currently being made to reuse such wastes.

Regarding strategies for recycling waste materials, and from a sustainability point of view, several authors mention that any innovation developed in different industrial sectors (production of materials and semi-finished products with recycled content) gives design solutions with attention to the environment, which represents a positive outcome [8]. Furthermore, Migliore et al. [8] point out that the efficient use of resources can contribute to economic growth and job creation, in addition to helping to offset the costs that the world's gross domestic product (GDP) should bear to keep global warming within the + 2 limit, respecting the limits set by the Paris agreement. Based on this background, the present project arises in a manufacturing company, presented below.

2.2 Case Study

The company in which the case study is carried out is located in Tijuana, Mexico. This company is of French origin and specializes in the production of aircraft interiors (seats, skies, windows, doors, bathrooms, monuments, and trunks) and aeronautical safety equipment. For the proper functioning of its operations, the company's structure is composed of eight elementary departments: Operations, Human Resources, Engineering, Quality, Operational Excellence, Program, Finance, and Materials.

The main activities of each department are listed below:

1. Operations: Responsible for the manufacture of products and ensuring that they are on time.
2. Human Resources: In charge of recruiting new talent, training them and verifying that the general processes comply with the standards of the different governmental agencies.
3. Engineering: In charge of the development of the design of new products and modifications to existing ones and the automation of processes.
4. Quality: In charge of conducting internal audits to validate that the processes and products meet the customers' expectations.
5. Operational Excellence: Responsible for continuous improvement, implementing methodologies that identify areas of opportunity concerning defects, waste, and downtime.

6. Program: The department is in frequent contact with the customer, periodically reviewing its demand to schedule production dates to deliver products on time.
7. Finance: It is in charge of payments to national and international suppliers, employees, and governmental agencies.
8. Materials: It is in charge of production planning, purchasing, and importing the inputs used for manufacturing.
9. Regarding the operations department, Fig. 2.1 shows the company's operational infrastructure, which comprises nine processes, starting with the Pressing and Rolling process and ending with the Assembly process. Table 2.1 describes what each of the processes consists of.

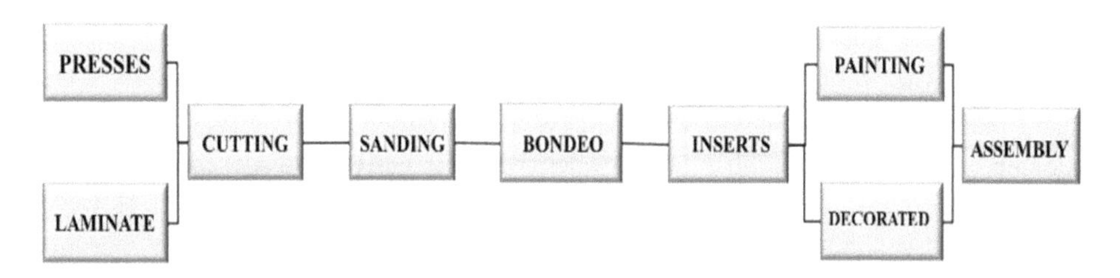

Fig. 2.1 Operational infrastructure of the company

Table 2.1 Production processes description

Process	Description
Mechanical press	Consists of placing the Core and pre-impregnated materials in different molds, transformed through a thermal process
Laminating	Consists of placing the Core and pre-impregnated materials in different molds to create the parts. After several layers of material, depending on the mold, the vacuum is performed to avoid leaving air bubbles
Cutting	The parts that come out of lamination or presses go to the cutting area to give the necessary finishing before the next process
Sanding	In this part, the pieces are polished to remove imperfections that may have been left in the pressing or laminating process
Bonding	It is in this area where the assembly of the pieces begins, which is done with different adhesives
Inserts	Subsequently, the part is cleaned of any excess adhesives that may have been left over from the previous process
Painting	The parts may require painting, which was carried out with a pneumatic gun
Decorating	Some other parts do not require painting but use a type of wallpaper placed with an adhesive on the entire part
Assembly	The required accessories or electrical assemblies are placed on the part

On the other hand, the Human Resources department is responsible for the Health and Safety area, where this case study is being conducted. This area is responsible for managing government permits and licenses, compliance with applicable regulations, monitoring internal and external audits, and implementing processes and safety measures to eliminate risks. It is also responsible for waste disposal.

The waste segregation system identified that the process is not standardized, so the project focuses on optimizing waste disposal costs by evaluating and analyzing the stages of generation, collection, storage, and disposal of special handling waste (SMW) and municipal solid waste (MSW).

2.2.1 Problem Statement

The project arose from the detection that waste management is not standardized, which means that there is a mixture of MSW, MSW and hazardous waste (HW) streams. One example is the disposal of fiberglass powder, which is currently disposed of as HWW. Currently, fiberglass powder costs 55 dollars/ton, plus the cost of transportation. However, its destination is confinement with general waste (which belongs to the MSW category) and costs 129 dollars/ton. Table 2.2 shows the fiberglass disposal costs in 2020, which totaled $15,751.45 for fiber disposal.

Regarding the disposal of garbage or MSW, it was found that the waste that is discarded can be recycled without generating any disposal costs. These wastes are wood, tinplate,

Table 2.2 Monthly cost of fiberglass disposal

Month	Cost (Dlls)	Tons
January	$1,972.85	42.8
February	$2,090.55	44.59
March	$1,756.70	49.82
April	$1,234.75	35.04
May	$1,102.75	27.1
June	$1,338.70	16.39
July	$1,377.75	25.12
August	$781.00	9.7
September	$779.35	27
October	$1,411.30	19.06
November	$1,070.30	12.35
December	$835.45	17.07
Total	$15,751.45	326.04

Table 2.3 Monthly cost of garbage disposal

Month	Cost (Dlls)	Tons
January	$8,815.29	90.74617
February	$7,481.19	61.10908
March	$9,318.50	66.84011
April	$4,475.73	35.04
May	$3,238.82	38.71808
June	$5,336.24	35.89409
July	$5,215.79	40.91441
August	$5,020.27	28.35842
September	$4,508.47	33.96499
October	$4,323.25	22.40858
November	$3,779.27	23.46764
December	$3,805.10	25.18673
Total	$65,317.91	502.6483

high-density plastic, and metal wrappings. Table 2.3 shows the costs for the disposal of these wastes.

In addition to the issue of MSW, it was detected that the segregation of MSW is not carried out effectively, which causes monetary losses at the time of sale of these wastes. The company's MSW includes aluminum, iron, and aluminum burrs; however, at the time of disposal, these wastes are mixed and disposed of as "metal", thus wasting their economic value. This is due to the lack of a waste separation system in the production areas. The purchase price of iron is 65 dollars/ton, that of aluminum burrs is 150 dollars/ton, and that of aluminum is 680 dollars/ton. Table 2.4 shows the monthly sales per metal in 2020.

As can be seen, the sum of waste disposal costs amounts to $81,069.36 ($15,751.45 for fiberglass +$65,317.91 for garbage disposal), while the income from metal sales amounts to $13,994.67, which results in a difference of 67,125.69 dollars in 2020. The above paragraphs highlight the need to carry out this project to standardize the waste segregation system, looking to reduce disposal costs and increase the income obtained from sales of special handling waste.

2.2.2 Research Objective

Based on the problem stated above, the objective of this case study is: To reduce waste disposal costs and increase waste sales by at least 15% using the Plan-Do-Check-Act (PDCA) cycle methodology.

Table 2.4 Monthly income from metal sales

Month	Cost (Dlls)	Tons
January	$2620.0525	40.3085
February	$2397.8825	36.8905
March	$1116.765	17.181
April	$692.9975	10.6615
May	$1024.985	15.769
June	$1148.68	17.672
July	$1431.885	22.029
August	$768.56	11.824
September	$1376.2775	21.1735
October	$435.6625	6.7025
November	$513.0125	7.8925
December	$416.91	6.414
Total	$13,943.67	214.518

2.3 Methodology

This section describes the methodology applied to achieve the objective planned in Sect. 1.2.2.

2.3.1 Materials

Table 2.5 below shows the material list used to carry out the project, as well as the quantities of each one of them. In addition to the materials mentioned above, the Excel program will be used to prepare reports regarding the disposition and follow-up of the activities.

The use of the excel sheet allowed the consecutive listing of the layout of all materials, which was easy to use and understand by all project participants. It is important to mention that initially, adjustments were made to the layout, as comments and suggestions were received from the participants.

2.3.2 Method

The method applied to carry out this project consists of applying the four phases of the PDCA cycle, and how each of them is executed is described below. Figure 2.2 shows the different phases and their respective steps in the method applied in this project.

Table 2.5 Materials used in the project

Aquntity	Material	Image
4	75 GAL containers. One of each color (Red, Green, Yellow and Blue)	
8	75 GAL containers. Two of each color (Red, Green, Yellow and Blue)	
3	Recycling stations. Each station has a red, green, yellow and blue container	
5	160 × 60 cm display stand	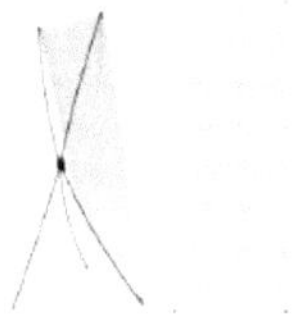
4	Sheet metal signs	
1	Industrial scale	

(continued)

Table 2.5 (continued)

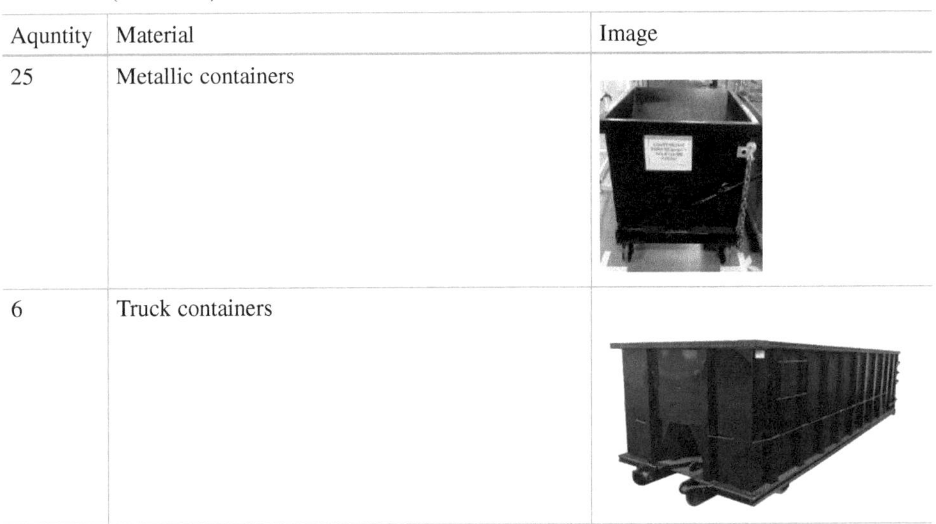

Aquntity	Material	Image
25	Metallic containers	
6	Truck containers	

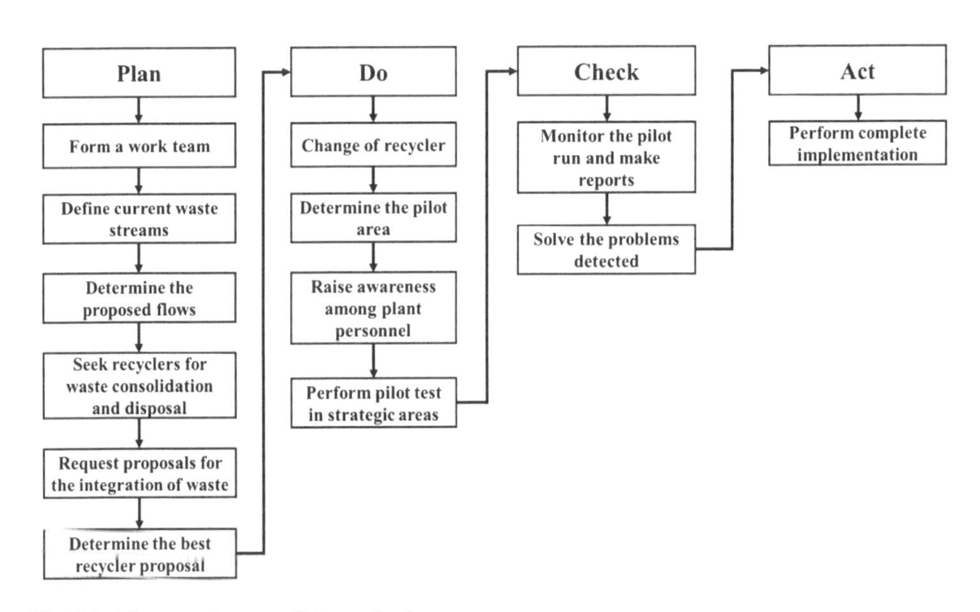

Fig. 2.2 Phases and steps of the method

2.3.2.1 Phase 1: Plan

In this phase, activities are planned, considering the objectives to be achieved and the tools needed are determined. Once the waste segregation project has been planned, a presentation is made to management for evaluation and acceptance. This phase is divided into five steps, which are described below.

Form the work team for the implementation of the project. In this step, a team is created, where team members must possess skills and knowledge that will contribute to the implementation of the project. In addition, they are assigned activities such as coordinating each stage of the project, coordinating the cleaning team, and defining who will be in charge of waste collection. Support personnel is also included in each stage of the project to design and communicate information about waste segregation. A person in charge of the waste collection area is also included.

Define the current waste streams. In this step, a survey is made throughout the plant to determine the types of waste streams (re-waste) that are handled in the different work areas of the plant, such as production areas, warehouses, administrative areas, restrooms, and canteens.

Determine the proposed flows. After detecting the different current waste streams in the company, each stream is segregated to determine which type of waste is to be disposed of in a single stream. Once the waste has been identified for each stream, samples of each waste are taken and provided to the recycling companies to determine if they can be recycled at no cost or taken at an economic value.

Seek out recycling companies for waste consolidation and waste disposal. One of the purposes of the project is to consolidate waste with a single recycling company for better waste management. For this reason, a search for recycling companies that can collect MSW and MSW is being conducted. The search for recycling companies is done through the purchasing department, which, through a bidding process, decides to select the candidate recycling companies based on: costs for disposal, the cost for the sale of MSW, and additional services.

Request proposals for waste integration. After determining the candidates for waste consolidation and disposal, each recycling company is given samples of the waste collected from the different streams (step 1). Based on this waste, the recyclers should put together their proposals for waste integration.

Determine the best proposal from the recyclers. After obtaining the proposals from the different recyclers, a bidding process is carried out, in which the following departments participate: Import and Export, Finance, Purchasing, and Health and Safety. A comparative table of the sale prices of recyclable waste and the cost of solid urban waste is drawn up during the bidding process. It is done to determine the best proposal from the recycling companies.

2.3.2.2 Phase 2: Do

In this phase, the "pilot" implementation of the project in a strategic area is initiated at scale. In addition, progress is monitored, and problems are corrected as they arise. To carry out this phase, the following steps are taken.

Change of recycling company. In the Plan phase, the best recycling company for waste integration is determined. After that, in this step, the following data is requested to register it in the catalog of recycling companies:

1. RFC/Taxpayer ID
2. Three commercial references
3. Opinion of compliance with tax obligations
4. Letterhead letter with the signature of the legal representative.
5. Bank details (Bank, account number, CLABE account)
6. Contact information (Name, position, telephone, and e-mail)
7. IMMEX registration certificate.

Once the new recycler is registered in the system, the change begins, notifying the current recycling companies that the service currently provided is no longer necessary due to the company's strategies. Once notified, they are given one week to remove their containers and bring in the new company's truck containers and metal containers. Within the area assigned to the new recycler, the industrial scale is installed to weigh the MSW, the sheet metal signs, and the 75-Gallon containers to identify which type of waste goes in which container.

Determine the pilot area. This step determines the areas where the project will be implemented. However, an area must be chosen for the pilot run. The area to be chosen should be the one that contains the highest waste generation and is an area with a high flow of personnel. It is intended to raise awareness among as many people as possible.

Raise awareness among plant personnel. Once the pilot area is determined, the 160 × 60 cm display is installed in the pilot area to inform plant personnel about the new changes in waste segregation. In addition to this, the information with the notification is sent via e-mail to all workers and administrative staff.

Conduct the pilot test in strategic areas. After determining the best area to conduct the pilot test, the recycling stations and metal containers are placed, accompanied by their respective signs.

2.3.2.3 Phase 3: Check

In this phase, the implementation of the pilot test of the project in the area determined in stage two is monitored. This test allows adjustments and adjustments to be made to the project. The follow-up consists of the following activities:

Monitor the pilot run and make reports. After conducting the pilot run, monitoring should be carried out:

Be in the area during the hours of peak personnel flow, explaining to the personnel the new way of segregating waste.

Ensure that personnel do not mix waste.

Share with personnel the importance of recycling.

Weigh the waste that is generated.

Make waste disposal reports.

Solve the problems detected. During the pilot test of the project, problems arise, which should be documented in the Excel® report and a list of actions taken to correct and prevent problems in the full implementation.

2.3.2.4 Phase 4: Act

In this last stage, the complete project is implemented throughout the plant. For this purpose, the following activity is performed:

Perform full implementation. After solving the problems resulting from the pilot test, the complete implementation is carried out in the areas, which involves placing the recycling stations with their signs and displays in each area.

2.4 Results

This section presents the results obtained by implementing the method described in the previous section. For a better understanding, these results are presented for each of the phases of the method described.

2.4.1 Findings from Phase 1: Plan

Based on the list of areas that would impact the implementation of the project, it was determined which workers would be part of the team to carry out the project. Table 2.6 shows the members selected to make up the work team and the area to which they belong, and the functions assigned to them. As can be seen, 3 workers were selected from the Health and Safety area, one from the Communication area, and another from the labor area.

Table 2.6 Members of the work team

Worker	Area	Function
HSE specialist	Safety and hygiene	Coordinate implementation activities
HSE technician	Safety and hygiene	Support in the different stages of the project
HSE stockist	Safety and hygiene	In charge of the waste collection area
Communication technician	Communication	Design and share information about the project
Service coordinator	Labor	Coordinate the team in charge of waste collection

Table 2.7 Different current waste streams detected at the plant

Stream name	Types of waste	Location
General trash	Sanitary waste Food waste Plastic containers (Soft drinks, Juices) Aluminum cans (Soft drinks, Juices) Food wraps (Sabritas, Cookies) Tin cans (Vegetables, tuna) Paperboard Wood Paper	Offices, toilets, in all production and warehouses
Fiberglass	Fiberglass powder Waste parts	CNC 5X, CNC 3X, Rolled
Metal	Solid aluminum Aluminum burr Iron	Machining, Deburring, CNC 3X, CNC 5X, Punching, Cutting

Similarly, Table 2.7 shows the names of the different current waste co-streams obtained at the company's plant and the areas in which they are located.

Concerning the search results for new recycling companies, four candidates were obtained, which were named Recycler1, Recycler1, Recycler3 and Recycler4. Table 2.8 shows the proposals of each recycler for the different wastes.

Based on the proposals provided by each recycler, the following was determined: the best cost for garbage and fiberglass disposal is provided by Recycler1, since it offers a maximum collection of four tons for USD 300, while the other recyclers charge per collection and weight. It can also be observed that, of the four proposals, only Recycler 1 offers to pay for aluminum cans, flexible packaging, tinplate and plastic waste, while the other companies offer to collect them free of charge. Table 2.9 compares the costs of each of the recyclers' proposals.

As can be seen, Recycler1 is the company that offers the lowest cost for garbage and fiberglass collection, charging 75 dollars per ton, while the other companies range from $98 to 131 dollars per ton. Table 2.10 shows the general waste and glass fiber disposal cost based on the number of tons generated in 2020.

Based on the results shown in Table 2.10, it was determined that Recycler1 handles the lowest costs for disposal of general waste and fiberglass. Table 2.11 shows the prices offered by each recycler for the purchase of recyclable waste such as cardboard, wood, aluminum, iron, aluminum burrs, aluminum cans, flexible packaging, tinplate, plastic, using an approximate amount of tons generated annually.

As can be seen in Table 2.11, the company that offered the highest profit for the sale of waste was Recycler1, offering approximately 14,300 dollars per year for the sale of waste, while the company that offered the lowest profit was Recycler2, offering 8,400 dollars

Table 2.8 Proposals from the recycling companies for the different types of waste

Waste name	Product value	Observations	Recycler
Industrial garbage	$300 USD	Maximum weight 4 tons ($30 dollars per additional ton)	Recycler1
	$175 USD	$175 per pickup + $55/ton. Maximum 4 tons per trip	Recycler2
	$140USD	$140 per pickup + $75/tonne. Maximum 4 tons per trip	Recycler3
	$125 USD	$125 per pickup + $100/ton. Maximum 4 tons per trip	Recycler4
Fiberglass	$300 USD	Maximum weight 4 tons ($30 dollars per additional ton)	Recycler1
	$175 USD	$175 per pickup + $55/ton. Maximum 4 tons per trip	Recycler2
	$140USD	$140 per pickup + $75/tonne. Maximum 4 tons per trip	Recycler3
	$125 USD	$125 per pickup + $100/ton. Maximum 4 tons per trip	Recycler4
Paperboard	$15 USD	Price per ton	Recycler1
	$20USD	Price per ton	Recycler2
	$25USD	Price per ton	Recycler3
	$2USD	Price per ton	Recycler4
Wood	$0USD	Free collection	Recycler1
	$0USD	Free collection	Recycler2
	$0USD	Free collection	Recycler3
	$0USD	Free collection	Recycler4
Aluminum	$700 USD	Price per ton	Recycler1
	$550 USD	Price per ton	Recycler2
	$700 USD	Price per ton	Recycler3
	$680USD	Price per ton	Recycler4
Long iron	$95 USD	Price per ton	Recycler1
	$120USD	Price per ton	Recycler2
	$75 USD	Price per ton	Recycler3
	$65USD	Price per ton	Recycler4
Aluminum burr	$220USD	Price per ton	Recycler1
	$150USD	Price per ton	Recycler2

(continued)

Table 2.8 (continued)

Waste name	Product value	Observations	Recycler
	$200USD	Price per ton	Recycler3
	$150USD	Price per ton	Recycler4
Aluminum cans	$5000 MXN	Price per ton	Recycler1
	$0 MXN	Free collection	Recycler2
	$0 MXN	Free collection	Recycler3
	$0 MXN	Free collection	Recycler4
Flexible packaging	$0 MXN	Free collection	Recycler1
	$0 MXN	Free collection	Recycler2
	$0 MXN	Free collection	Recycler3
	$0 MXN	Free collection	Recycler4
Tin	$1500 MXN	Price per ton	Recycler1
	$0 MXN	Free collection	Recycler2
	$0 MXN	Free collection	Recycler3
	$0 MXN	Free collection	Recycler4
Plastic	$1500 MXN	Price per ton	Recycler1
	$0 MXN	Free collection	Recycler2
	$0 MXN	Free collection	Recycler3
	$0 MXN	Free collection	Recycler4

USD = American dollars, MXN = Mexican pesos

Table 2.9 Comparison of prices in dollars by each of the recyclers

Type of waste	Recycler1	Recycler2	Recycler3	Recycler4
Industrial garbage	$75.00	$98.75	$110.00	$131.25
Fiberglass	$75.00	$98.75	$110.00	$131.20
Cardboard	$15.00	$20.00	$25.00	$2.00
Wood	$0	$0	$0	$0
Aluminum	$700.00	$550.00	$700.00	$680.00
Long iron	$95.00	$120.00	$75.00	$65.00
Aluminum burr	$220.00	$150.00	$200.00	$150.00
Aluminum cans	$250.00	$0	$0	$0
Tin	$75.00	$0	$0	$0

Table 2.10 Comparison of cost in dollars for the disposal of MSW per recycler

Type of waste	Tonnes generated in 2020	Recycler1	Recycler2	Recycler3	Recycler4
Industrial garbage	326	$24,450.00	$32,192.50	$35,860.00	$42,787.50
Fiberglass	260	$19,500.00	$25,675.00	$28,600.00	$34,112.00
Cost		$43,950.00	$57,867.50	$64,460.00	$76,899.50

Table 2.11 Prices in dollars for the purchase of recyclable waste by recyclers

Type of waste	Annual tons	Recycler1	Recycler2	Recycler3	Recycler4
Paperboard	10	$150.00	$200.00	$250.00	$20.00
Wood	10	$-	$-	$-	$-
Aluminum	10	$7,000.00	$5,500.00	$7,000.00	$6,800.00
Long Iron	10	$950.00	$1,200.00	$750.00	$650.00
Aluminum Burr	10	$2,200.00	$1,500.00	$2,000.00	$1,500.00
Aluminum Cans	Two	$2,500.00	$-	$-	$-
Flexible Packaging	Two	$-	$-	$-	$-
Tin	Two	$750.00	$-	$-	$-
Plastic	Two	$750.00	$-	$-	$-
Sales		$14,300.00	$8,400.00	$10,000.00	$8,970.00

per year, 41% less than Recycler1. Based on these results, it was determined that the best company, both for waste disposal and the sale of waste, was the company Recicladora1, with an estimated profit of 14,300 dollars for the disposal of special handling waste and an expense of 43,950 dollars for the disposal of urban solid waste.

2.4.2 Findings from Phase 2: Do

After selecting the recycler1, the company delivered five containers for trucks, which were distributed as follows: two for general waste, one for aluminum and long iron waste, one for cardboard and one for fiberglass. In addition to this, three metal holders were placed for the aluminum burr. Figure 2.3 shows these holders.

The area was chosen as the "pilot area" for the segregation of flexible packaging, aluminum cans, plastic and tinplate were the canteen area of the plant since it is an area where these types of waste are generated and where there is a greater flow of personnel. For the segregation of the different metals, the "pilot area" was the die-cutting area since its production process uses different metals such as aluminum and iron, generating aluminum, aluminum, and iron burr waste. The communication area was designed with nine

Fig. 2.3 Recycler1 containers

different signs, which were placed in key areas of the plant and the areas chosen as pilot areas. Figure 2.4 shows two of these sign designs.

Figure 2.5a shows the containers placed in the die-cutting area designated for aluminum and iron waste. Figure 2.5b shows the recycling station used in the canteen area to segregate recyclable waste.

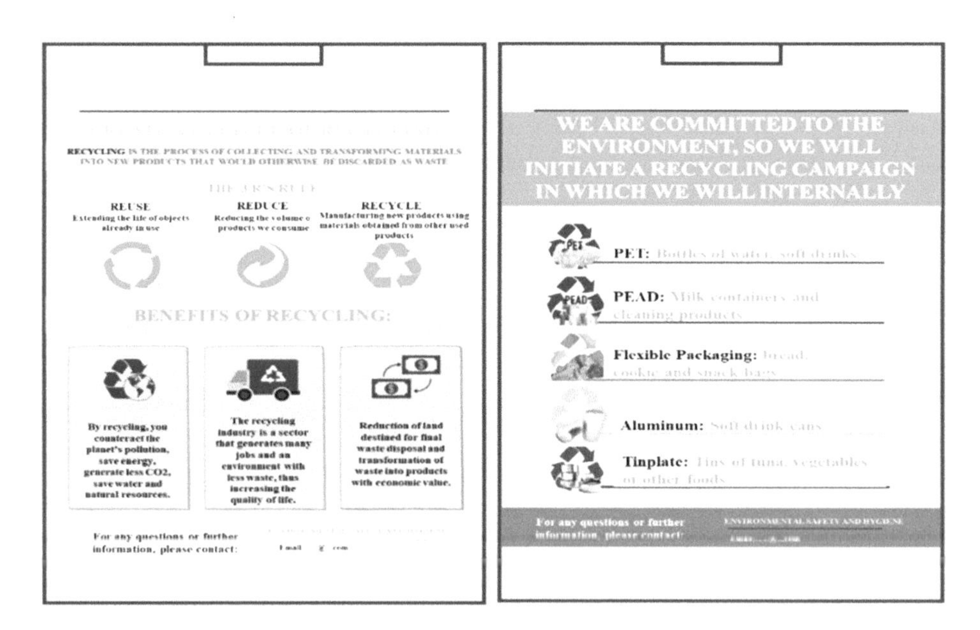

Fig. 2.4 Samples of signs used to raise awareness of recycling

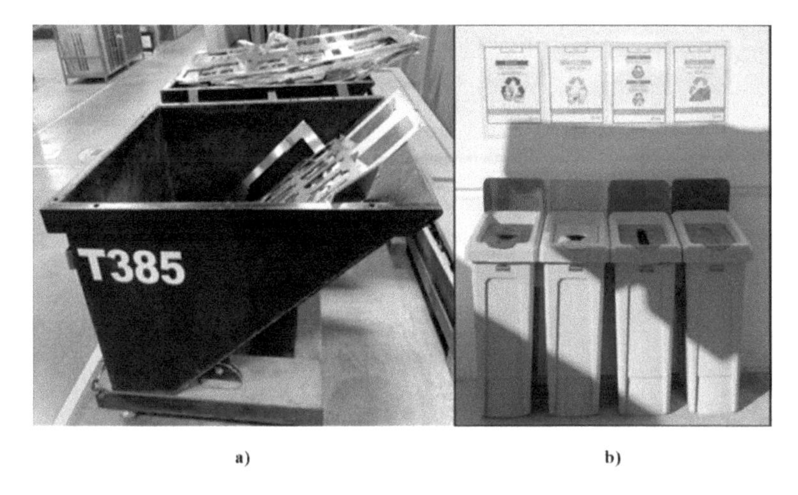

Fig. 2.5 Containers: **a**) for the die-cutting area, **b**) for the lunch area

2.4.3 Findings from Phase 3: Check

The monitoring of the pilot run was carried out for 30 days in August 2020. Table 2.12 shows the waste generated during this period, in which the waste generated was a general waste, aluminum burrs, long iron, aluminum, tinplate, aluminum cans, flexible packaging and plastic.

As can be seen, 41.61 tons of waste were generated, of which 35.98 tons corresponded to general waste. The expectation was 10% reusable waste, which is equivalent to 3500 kg of waste. On the other hand, 710 kg of tinplate, aluminum cans, flexible packaging and plastic were generated. Finally, 400 kg of aluminum, 2710 kg of aluminum burrs, and 1810 kg of long iron were obtained.

During the pilot run, some problems arose that prevented the project from achieving its goal of reducing general waste by 10%, with only a 1% reduction. The root cause of this problem was identified as a lack of awareness and discipline among the plant's employees. It was because, when depositing their waste in the containers, they deposited everything in a single container, without respecting the signage, since, having different types of containers, they mixed the garbage or threw common garbage in the recycling bins.

This problem was solved by placing a resource from the cleaning area and separating what had already been mixed. To correct and avoid repetitions, larger signs were placed to make them more visible, and people monitored when the workers were depositing their waste in the garbage cans to make them aware that they were doing it incorrectly. This created awareness among the company's workers.

Table 2.12 Waste generated during the pilot run

Residue	Containers	Actual weight (Kg)
Trash	Roll-off	3130
Trash	Roll-off	3020
Trash	Roll-off	2990
Trash	Roll-off	3040
Aluminum burr	Roll-off	1520
Trash	Roll-off	2370
Trash	Roll-off	2550
Trash	Roll-off	2610
Trash	Roll-off	2520
Trash	Roll-off	2730
Aluminum	Roll-off	400
Trash	Roll-off	2450
Trash	Roll-off	3460
Aluminum burr	Roll-off	1190
Trash	Roll-off	2260
Trash	Roll-off	2700
Long iron	Roll-off	1810
Trash	Roll-off	150
Tin	Roll-off	100
Aluminum cans	Roll-off	230
Flexible packaging	Roll-off	30
Plastic	Roll-off	350
Total		41,610

2.4.4 Findings from Phase 4: Act

Once the problems presented during the pilot run were solved, the next step was to place aluminum, iron, and aluminum waste containers in the machining, 5-axis and deburring areas. Recycling stations for separating tinplate, aluminum cans, flexible packaging, and plastic waste was installed in the office, terrace, and tent areas. An example is shown in Fig. 2.6.

2.4.5 General Findings

Table 2.13 shows the list of waste generated from August through November 2021.

Fig. 2.6 Recycling station (Terrace)

With the waste generated from August to November 2020, the sale of MRE was USD 12,046.74 with Recycler1. This was because waste segregation was carried out, and other types of waste were added to sell these, such as tinplate, aluminum cans, flexible packaging, and plastic. Figure 2.7 shows the income comparison from the sale of waste before and after the project. As can be seen, income from sales increased significantly.

Likewise, Fig. 2.8 shows the comparison of general waste disposal costs, which decreased by 48.45%, achieved by reducing the cost of collection and segregating waste with a commercial value. As can be seen, costs decreased significantly, with more than USD 3,000 per month savings.

Finally, Fig. 2.9 shows the cost comparison for fiberglass disposal, which was reduced by 70.59%. This was because the cost per ton of fiberglass was 55 dollars, and its transportation cost 200 dollars. Currently, only 75 dollars per ton is charged, with no transportation cost.

2.5 Conclusions

The general purpose of the project was to reduce the cost of waste disposal generated in the various work areas using the PDCA methodology. This objective was achieved satisfactorily since, in the case of waste disposal costs, these were reduced by 48.45%, while in the case of fiber, the reduction was 70.59%, resulting in an overall reduction of 56.93%. In addition to this, an 81% increase in sales of special handling waste was achieved.

Based on this, it is concluded that the PDCA methodology is a useful tool not only for solving product quality problems, as stated by authors such as Realyvásquez-Vargas et al. [9], Nabiilah et al. [10] or Rahman et al. [11], or in saving energy and delivery

Table 2.13 Waste generated from August to November 2020

Month	Residue	Actual weight (kg)
August	Aluminum	400
	Trash	35,980
	Flexible packaging	30
	Long iron	1810
	Tin	100
	Aluminum cans	230
	Plastic	350
	Aluminum burr	2710
September	Aluminum	2720
	Trash	58,410
	Paperboard	5440
	Fiberglass	20,180
	Long iron	9210
	Wood	6700
	Aluminum burr	2010
October	Aluminum	1110
	Trash	44,870
	Paperboard	7200
	Fiberglass	15,060
	Long iron	5020
	Wood	1650
	Aluminum burr	2260
November	Aluminum	4000
	Trash	53,439
	Paperboard	5940
	Fiberglass	8490
	Long iron	1960
	Wood	0
	Aluminum burr	1980
	Plastic	350

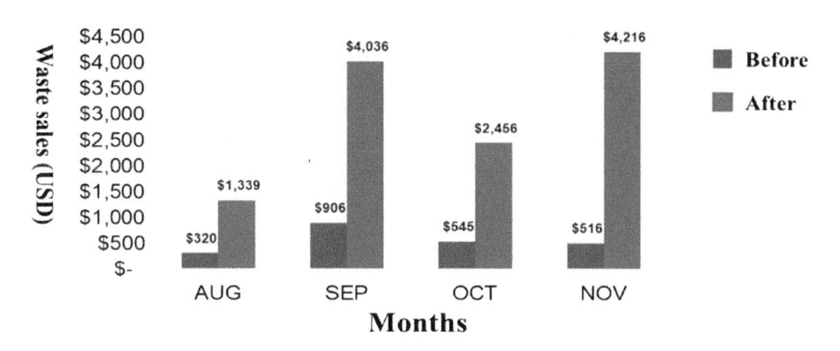

Fig. 2.7 Comparison of RME sales revenues

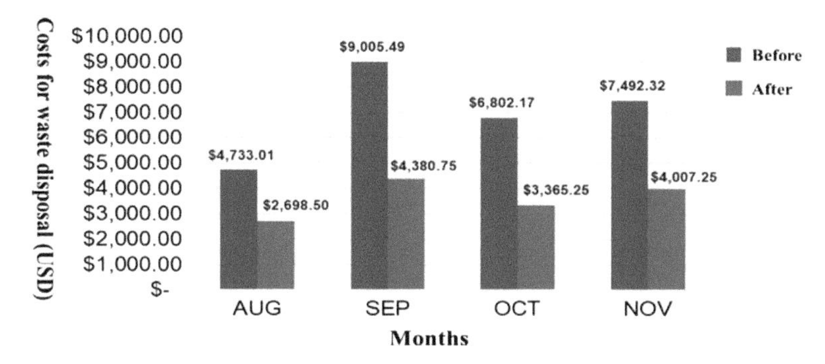

Fig. 2.8 Comparison of costs for waste disposal

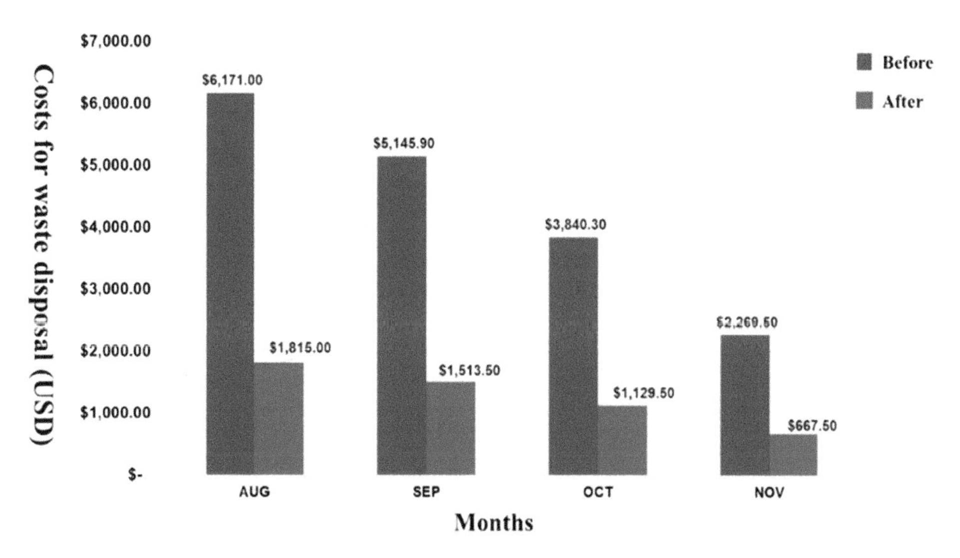

Fig. 2.9 Comparison of costs for fiberglass disposal

times [12], but it is also useful for solving resource management problems, in this case, waste. In all these cases, the PDCA cycle contributes to increase process quality. In the specific case of this research, it is concluded that it fulfills what has been mentioned by several authors concerning the fact that this method allows continuous, incremental, fast and effective improvements without making large investments [12–18].

Proper waste segregation generates social benefits, such as employment and a cleaner environment and, consequently, an increase in the quality of life. It also generates economic benefits, as land for waste disposal is reduced, and waste is transformed into products with economic value. Finally, correct waste segregation also generates environmental benefits since recycling counteracts global pollution, saves energy, generates less CO_2, and saves water and natural resources.

However, a key element in this project was the workers' awareness of recycling, as they realized that small actions, such as depositing waste in the correct container, have a great impact. Therefore, it is concluded that the PDCA cycle is a tool that can help in the sustainability of resources in the manufacturing industry and the sustainability of the environment, an issue of great importance today.

References

1. Singh S, Ramakrishna S, Gupta MK (2017) Towards zero waste manufacturing: a multidisciplinary review. J Clean Prod 1(168):1230–1243
2. Rybicka J, Tiwari A, Alvarez Del Campo P, Howarth J (2015) Capturing composites manufacturing waste flows through process mapping. J Clean Prod 15(91):251–261
3. Shahbazi S, Salloum M, Kurdve M, Wiktorsson M (2017) Material efficiency measurement: empirical investigation of manufacturing industry. Procedia Manuf 1(8):112–120
4. Jain P, Gupta C (2018) Finding treasure out of textile trash generated by garment manufacturing units in Delhi/ NCR. GARI Int J Multidiscip Res 4(4):40–60
5. Mativenga PT, Agwa-Ejon J, Mbohwa C, Sultan AAM, Shuaib NA (2017) Circular economy ownership models: a view from South Africa industry. Procedia Manuf 1(8):284–291
6. Nilakantan G, Nutt S (2015) Reuse and upcycling of aerospace prepreg scrap and waste. Reinf Plast 59(1):44–51
7. Nilakantan G, Nutt S (2017) Reuse and upcycling of thermoset prepreg scrap: case study with out-of-autoclave carbon fiber/epoxy prepreg. J Compos Mater 52(3):341–60 [cited 2022 Mar 22]. https://doi.org/10.1177/0021998317707253
8. Migliore M, Carpinella M, Paganin G, Paolieri F, Talamo C (2018) Innovative use of scrap and waste deriving from the stone and the construction sector for the manufacturing of bricks. Environ Eng Manag J [Internet] 17(10):2507–14 [cited 2022 Mar 22]. http://www.eemj.icpm.tuiasi.ro/pdfs/vol17/full/no10/23_111_Migliore_18.pdf
9. Realyvásquez-Vargas A, Arredondo-Soto KC, Carrillo-Gutiérrez T, Ravelo G (2018) Applying the Plan-Do-Check-Act (PDCA) cycle to reduce the defects in the manufacturing industry a case study. Appl Sci 8(11):1–17
10. Nabiilah AR, Hamedon Z, Faiz MT (2018) Improving quality of light commercial vehicle using PDCA approach. J Adv Manuf Technol [Internet] 12(1(2)):525–34 [cited 2018 Oct 29]. http://journal.utem.edu.my/index.php/jamt/article/view/4310

11. Rahman M, Dey K, Kapuria TK, Tahiduzzaman M (2018) Minimization of sewing defects of an apparel industry in Bangladesh with 5S & PDCA. Am J Ind Eng [Internet] 5(1):17–24 [cited 2018 Oct 29]. http://pubs.sciepub.com/ajie/5/1/3
12. Isniah S, Purba HH, Debora F (2020) Plan do check action (PDCA) method: literature review and research issues. J Sist dan Manaj Ind [Internet] 4(1):72–81 [cited 2022 Feb 15]. https://e-jurnal.lppmunsera.org/index.php/JSMI/article/view/2186
13. Schneider PD (1997) FOCUS-PDCA ensures continuous quality improvement in the outpatient setting. In: Oncology nursing forum, p 966
14. Kholif AM, Abou El Hassan DS, Khorshid MA, Elsherpieny EA, Olafadehan OA (2018) Implementation of model for improvement (PDCA-cycle) in dairy laboratories. J Food Saf [Internet] 38(3):e12451 [cited 2018 Oct 29]. https://doi.org/10.1111/jfs.12451
15. Muhammad S (2015) Quality improvement of fan manufacturing industry by using basic seven tools of quality: A case study. Int J Eng Res Appl 5(4):30–35
16. Sangpikul A (2017) Implementing academic service learning and the PDCA cycle in a marketing course: contributions to three beneficiaries. J Hosp Leis Sport Tour Educ [Internet] 21:83–7 [cited 2018 Oct 5]. https://www.sciencedirect.com/science/article/pii/S1473837617300412
17. Silva AS, Medeiros CF, Vieira RK (2017) Cleaner production and PDCA cycle: practical application for reducing the cans loss index in a beverage company. J Clean Prod [Internet] 150:324–38 [cited 2018 Oct 10]. https://www.sciencedirect.com/science/article/pii/S0959652617304687
18. Maruta R (2012) Maximizing knowledge work productivity: a time constrained and activity visualized PDCA cycle. Knowl Process Manag 19(4):203–14 [cited 2018 Oct 10]. https://doi.org/10.1002/kpm.1396

3 Case Study 2. Raw Material Receipt Process Optimization

3.1 Introduction

In today's high global competitiveness, manufacturing companies are forced to improve the efficiency of all their processes [1]. One of the processes that have a major impact on competitiveness is the raw material procurement process [2], and it is one of the key elements of the efficiency of the material flow along the supply chain [3]. Its importance is evidenced by the fact that, for example, automobile manufacturers typically spend more than 50% of their sales on purchasing goods and services [4]. However, the efficiency of this process depends not only on the company that receives the raw materials but also on who supplies them [5]. Costa et al. [6] declare that the supply chain management process involves coordinating and collaborating activities among its different actors. Therefore, problems can occur during this process, such as late delivery by the supplier, material received on time but of poor quality, relatively long process times, and costs [7].

The raw material delivery problem has been the subject of research and scientific interest; for example, Mahendrawathi et al. [8] applied process mining to analyze the receipt of materials and found that bottlenecks existed, as the material spent too much time on the shelf. In addition, there was a deviation from the standard procedure established by the company, as additional activities were added, as well as a non-standard flow of materials, and the root cause of these problems was insufficient warehouse capacity.

Similarly, Costa et al. [6] report the case of a company that faced a problem related to the lack of traceability of its raw materials, which in turn led to negative consequences, such as a long response time to its customers and fulfilling an order, as well as a higher scrap cost due to the end of life of the raw materials.

Bradford and Gerard [9] document the case of a company that manufactured office furniture, which had experienced problems in its raw material purchasing process, such as long cycle times in the purchasing process, poor quality materials from its suppliers,

A. Realyvásquez Vargas et al., *The PDCA Cycle for Industrial Improvement*, Synthesis Lectures on Engineering, Science, and Technology,
https://doi.org/10.1007/978-3-031-26805-2_3

as well as dissatisfaction from its customers, since the longer it took for raw materials to arrive at the company and the longer it took for customers to receive their orders.

For their part, Fauzan et al. [10] mention the case of a company facing different problems related to the process of receiving raw materials, such as the lack of an automatic calculation of the resources held by the company, the process was not maximized to regulate the input and output of goods, there were no reports on inventory and expenditure incurred, and there is no computerized data storage.

Wutthisirisart et al. [11] point out that due to increased demand, companies often need more storage space, and often, the option of building a warehouse is not viable due to its high initial investment cost, which results in excess inventory that cannot be stored in the company's own warehouses being moved to third-party warehouses. This results in the company's rental payments, labor, and transportation costs, as it is forced to store the items and move them back to the production center.

Also, Frontoni et al. [12] mention that procurement and logistics organization are complex tasks in companies with multiple production/storage centers and indicate that these tasks become more difficult when there is an increase in demand variability and a frequent rotation of trends, which causes products to quickly become outdated.

Finally, Cortinhal et al. [13] indicate that the design of the supply chain network determines how many and which suppliers should be selected, in addition to the distribution channels through which materials flow from suppliers to customers, as well as the location and capabilities of the facilities to be operated. For this reason, designing the sourcing process represents a key element in improving a company's financial results since a successful supply network design sets the conditions for reducing costs and ensuring the operational efficiency of supply chain-related functions such as sourcing, production and distribution.

As seen in previous studies, even today, manufacturing companies face raw material supply problems, especially in a globalized production system, as the company presented in the following sections.

3.2 Case Study

The case study presented in this chapter is conducted in a manufacturing company that manufactures electronic products. Ninety-two percent of the company's operations are focused on manufacturing electronic components for the automotive industry, with mostly automated operations. With this, the company seeks to be a leader in the management of technology, systems and process control.

Its operational infrastructure is classified into two types of operations, Surface Mount Technology (SMT) and Manual Assembly; where the first type of operations is distributed in 11 automated lines, mostly made up of the following machines: PCB cleaner, printer, SPI (Solder Paste Inspection), Mounting Machines, Reflow Oven and AOI (Auto Optical

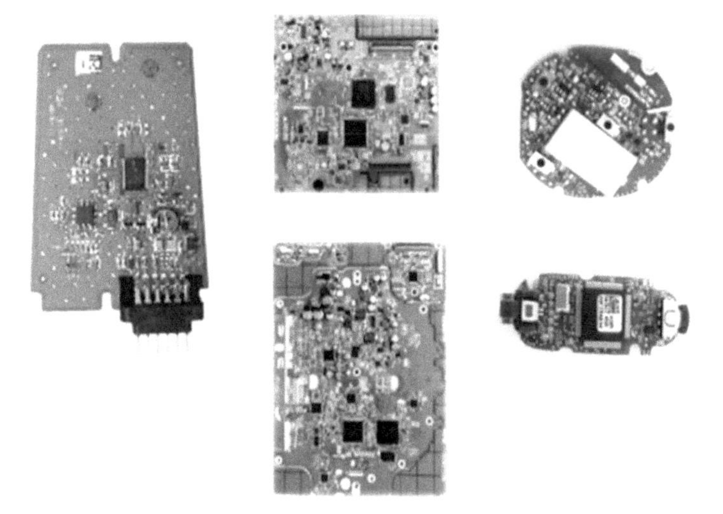

Fig. 3.1 Some of the products manufactured by the company

Inspection), while the second type of operations has approximately 13 lines, divided into operations such as conformal placement, In-circuit and Function Test (ICT), component assembly and packaging.

The above processes are part of the flow of most of the products, and their production is supervised by personnel certified in Acceptability of Electronic Products (IPC), from the inspector level to the management level in order to ensure product quality, which is mainly printed circuit boards (PCB's) for audio, sound, rear camera control, headlights and taillights of different ranges. Its main automotive customers are companies such as Mitsubishi Electric, Mazda, Nissan and Honda, and Plantronics in generic products. Figure 3.1 shows some of the products manufactured by the company.

The company's organizational structure is made up, of the first level, of the Presidency, followed by the divisions of Administration and Finance, Program Management or Business Group, and Plant Management. Subsequently, the different departments that make up the operational part of the company, such as Engineering, Manufacturing, Materials Control and Quality Planning and Assurance, are broken down. These departments are divided into more specific areas, as shown in Fig. 3.2.

This study is carried out in the Warehouse area, part of the Materials Control and Planning department. Figure 3.3 shows the organizational chart of the Warehouse area, and the functions of each of the workers in this area are mentioned below.

Materials Manager: Ensures the effective functioning of the materials process, coordinates activities of the different areas under his responsibility, clearly defines the metrics of the materials department and monitors the scope of these metrics.

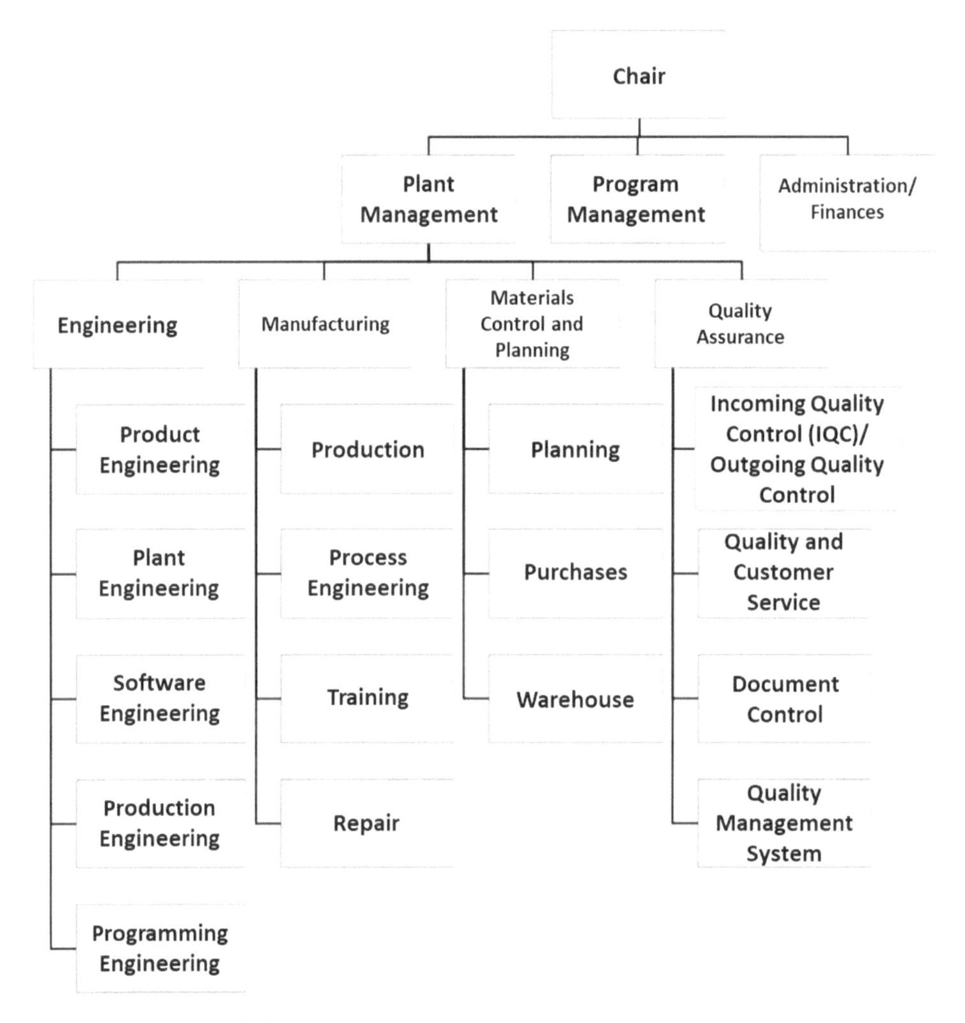

Fig. 3.2 Organizational chart of the company

General Materials Supervisor: Coordinates and supervises the activities derived from the functions of the department that contribute to the optimization of material and human resources.

Buyer: Procures supplies and materials according to specifications. Collaborates with other departments to determine supply requirements.

Planner: Plans production following management guidelines. Designs and directs production plans and material requirements. In addition, determines and calculates production volumes in coordination with Program Managers.

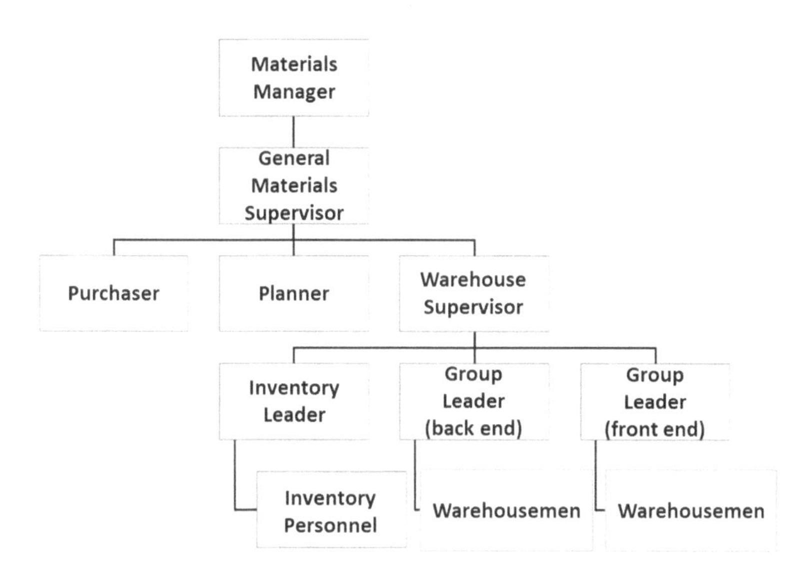

Fig. 3.3 Organizational diagram of the warehouse area

Warehouse Supervisor: Coordinates and supervises the procedures for receiving and storing materials.

Inventory Leader: Manages the inventory control system, maintains updated information, rectifies calculation errors, reviews and analyzes the results of operations.

Group Leader: Oversees that warehouse department operations are performed following operating instructions.

Inventory personnel: Performs physical inventories of materials.

Warehouseman: Follows the operating instructions of the warehouse department.

Figures 3.4 and 3.5 show the distribution of the company and the Warehouse area in which the study is carried out, respectively. As can be seen, the warehouse is divided into six sub-areas: Receipts, Shipping, Labeling, Finish goods, Back End and Front End. In the Back End sub-area, the processes of receipts, unpacking, shipments and storage of finished product are carried out, while in the Front End sub-area, the storage of raw materials and labeling of materials is located. Each of the processes is explained below.

Receipts: This is the process by which the company receives raw materials from suppliers. The transport is received, and the material is unloaded in the Receiving sub-area, and it is here where it is confirmed that the material complies with the documents.

Unpacking: This is the collection of material to be labeled. The packaging is removed and grouped by part number to be transferred to the labeling sub-area.

Shipments: This is the process of preparing finished product shipments to customers.

Finished product storage: In this process, the finished product is received from the production department, known as Finish Good, and stored in the finished product area of the Warehouse, ready to be shipped to the customer.

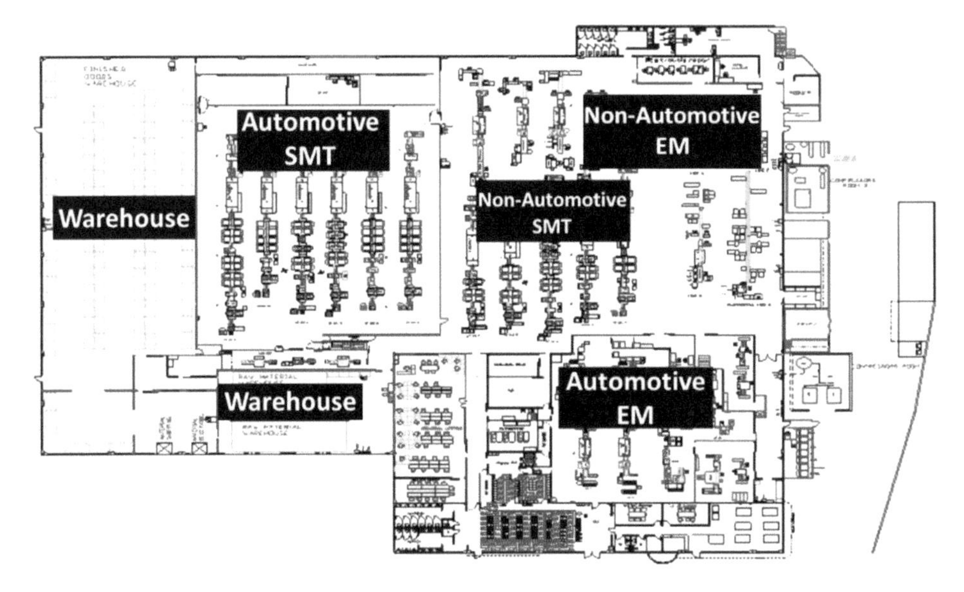

Fig. 3.4 Company's plant layout

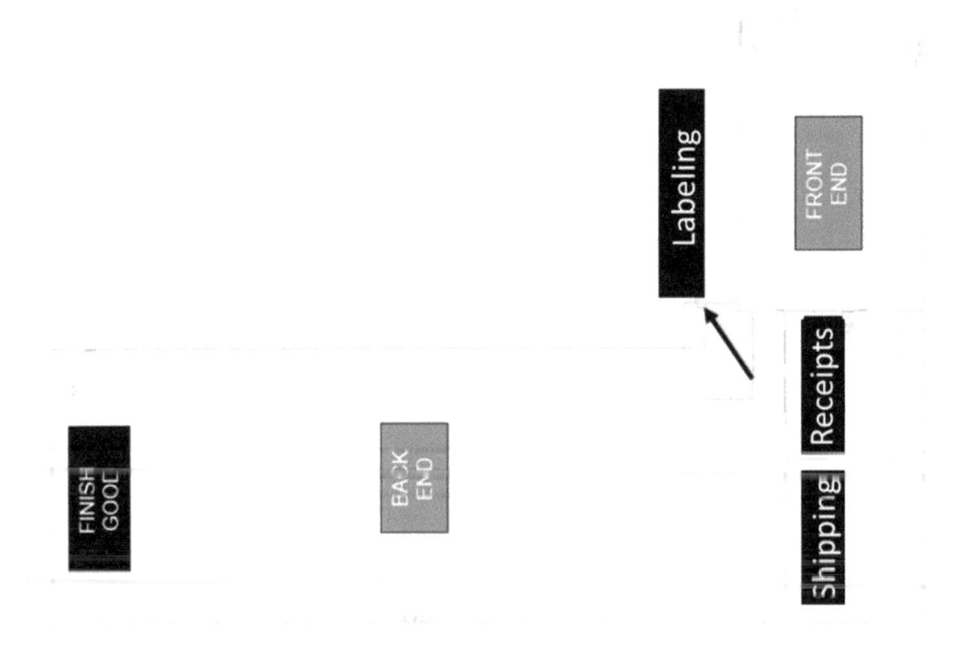

Fig. 3.5 Distribution of the warehouse area

Material labeling: This is the placement of a label with a unique bar code on each material received, and its traceability is recorded in the company's production process. Labeling can be automatic if the material has a barcode printed on the supplier's information or manual if it does not have a printed barcode.

Raw material storage: This is storing all raw material already labeled in its assigned location and ready to be delivered to production.

Three types of material are received and stored at the company's facilities, a description of which is given in Table 3.1.

3.2.1 Problem Statement

The company's warehouse receives up to 15 pallets of raw materials daily, which are imported mainly from Asia and the United States of America. All the raw material received goes through a production process, which is transformed into a finished product and sent to the client. One of the first activities in the production process is to receive the material, which goes through an inspection until it is placed in its assigned location for storage, ready to be delivered when required by the production area. Generally, receiving the material consists of the operations shown in Fig. 3.6.

Table 3.2 shows the capacity of units processed per shift and the total number of people per operation. The capacity for receiving raw material is currently unknown since the operation is carried out only once per import, which is identified with a consecutive ID number, and each invoice within the import is tracked to determine the time (in days) it is received and stored at its location.

It is observed that the labeling operation has the lowest capacity, representing a bottleneck that generates the accumulation of in-process inventory, and therefore, this project addresses that problem.

As part of continuous improvement, the company has observed that the time it takes to receive materials is 45 days, as seen in Fig. 3.7, with data collected from August to October 2021.

On the other hand, the lack of storage space in the Receiving sub-area and a stoppage of operations during April and May caused an excess of material inventory. This was because there was no production, but the receipt of material did not stop, forcing them to work overtime. Figure 3.8 shows the behavior of the expense incurred for overtime required by the Warehouse area, exclusively for material receipt operations during 2021 (January to October 2021).

In general, and in summary, the following problems have been encountered in the material receipt process:

Table 3.1 Types of materials stored by the company

Image of the material	Description
Front End	
	Rolls of electronic and semiconductor components
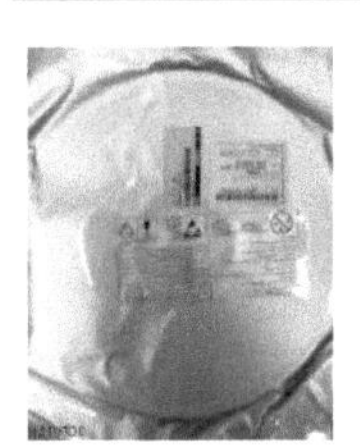	Rolls of electronic and semiconductor components with an antistatic bag
	Electronic components trays
	Electronic component trays with an antistatic bag
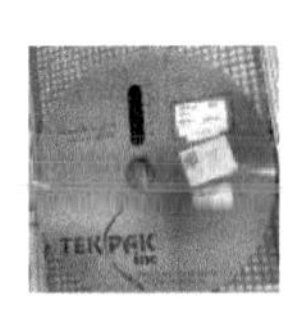	Rolls of metal components

(continued)

Table 3.1 (continued)

Image of the material	Description
Back End	
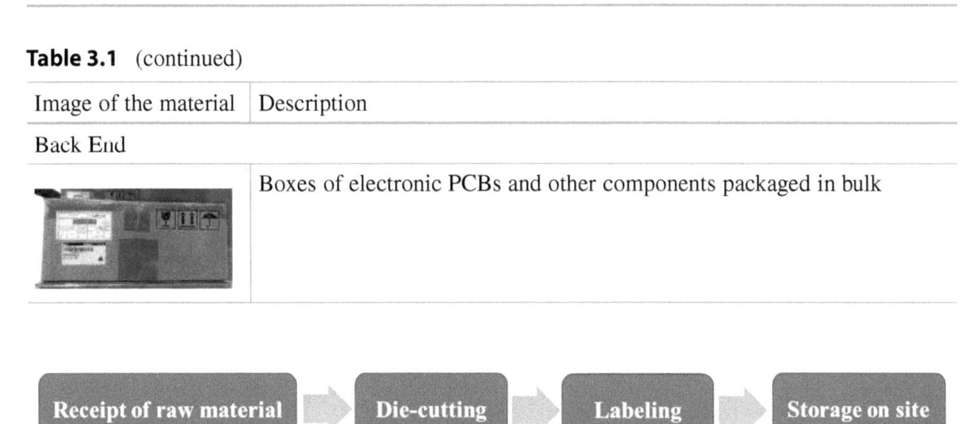	Boxes of electronic PCBs and other components packaged in bulk

Receipt of raw material → Die-cutting → Labeling → Storage on site

Fig. 3.6 Operations of the raw material receiving process

Table 3.2 Capacity of units processed per operation (per day)

Operation	Capacity (units)	Operators (3 shifts)
Receipt of raw material		
Die-cutting	1,820	
Tagged	580	
On-site warehousing	1,720	
Total	4,120	

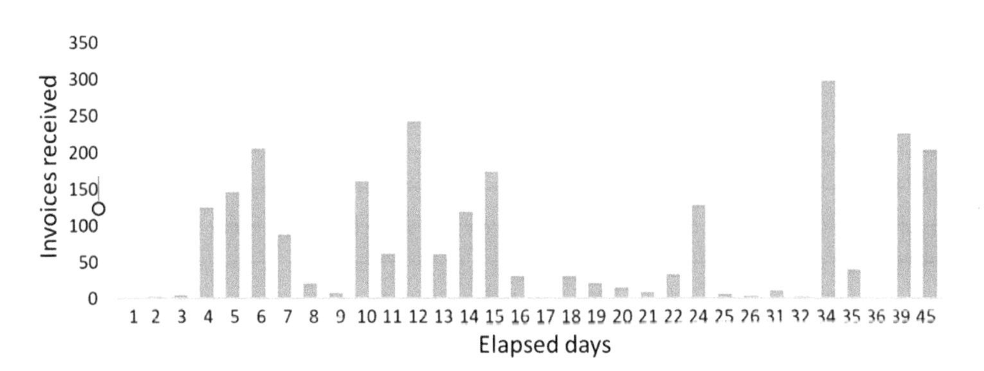

Fig. 3.7 Days elapsed in completing receipt of materials by invoices

1. It takes up to 45 days to complete the raw material storage process.
2. There is no storage space available in the receiving area.
3. It is necessary to work overtime to move forward with the receipt of materials.

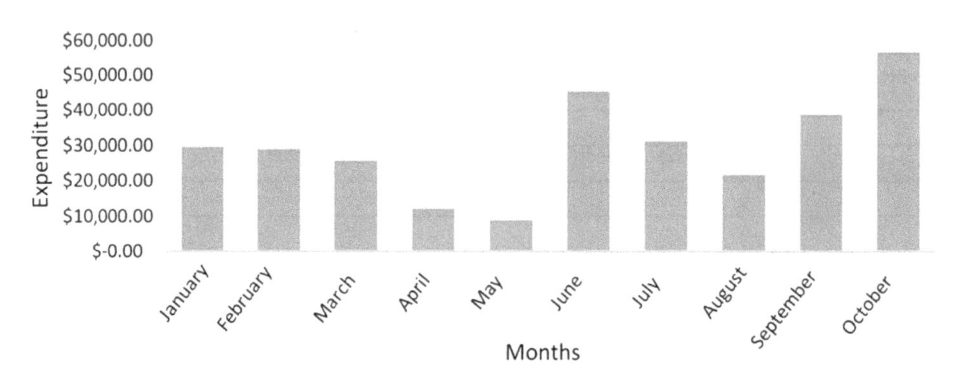

Fig. 3.8 Monthly trend overtime

3.2.2 Research Objectives

3.2.2.1 General Objective

The overall objective of this case study is to reduce the material receiving process time.

3.2.2.2 Specific Objectives

The specific objectives of the project are as follows:

1. Increase the labeling operation by at least 50% of the capacity of processed units.
2. Reduce 20% of the activities of the raw material reception process.
3. Eliminate 100% of the overtime required for the reception of raw materials.

3.3 Methodology

This section describes the methodology applied to achieve the objectives mentioned above.

3.3.1 Materials

The materials, tools and programs used during the project are as follows:

- Pallets
- Pallet Jack
- Workstations
- Scanner

- Computer equipment
- Tags
- Import documents
- Format for analysis of operations
- Microsoft Office Excel®
- Tool for identification and delimitation of work areas.
- Software systems.

3.3.2 Method

The tool applied to carry out this project is the PDCA cycle, which consists of 4 stages: Plan, Do, Check and Act. Next, it is explained in detail how they are applied to the project's development.

3.3.2.1 Step 1. Plan

In this first stage, the following general project data are defined:

- Project name
- Company
- The business line
- General and specific objectives
- Justification
- Delimitation and scope.

In addition, the members of the work team are also identified. Subsequently, the raw material reception task is analyzed through different diagrams. For example, the flow process chart of raw material reception is made to know the different subtasks that compose it.

After a general visualization of the current state of the process, a cause and effect diagram (also known as an Ishikawa diagram or fish diagram) is made. This is done to define the main problems, as well as their possible causes, focused on answering the question of why the process of receiving materials takes up to 45 days to complete. To carry out it, a team of workers involved in the process was assembled, including the General Materials Supervisor, the Warehouse Supervisor, warehouse leaders and storekeepers.

On the other hand, in order to solve as many of the problems detected as possible and to increase efficiency and optimize the process, a multidisciplinary work team is formed (general materials supervisor, warehouse supervisor, planner, general systems supervisor, systems engineer, and process engineer) to search for the best solutions to eliminate the main problems encountered. After a meeting, where the information obtained up to that moment is explained to the team members, a brainstorming session is held to define the actions to be implemented, and a schedule for implementing these actions is drawn up.

3.3.2.2 Stage 2. Do

At this point, the main problems that gave rise to this project must have been identified. Once this is done, the actions proposed in the previous stage to solve these problems are implemented. In addition, resources and people in charge are also designated to finally carry them out. Using the flow process chart made in the previous stage as a reference, the operations of the raw material receipt process are analyzed and, taking into account the implemented actions; a conclusion is obtained for each operation regarding whether it is necessary or not. Those that turn out not to be necessary are eliminated.

Subsequently, the 5's tool is implemented as follows:

- Eliminate: Here, you identify and separate materials and components necessary and unnecessary for performing tasks at each workstation.
- Order: The items identified as necessary are then organized, and a place is defined for their location to be easily found.
- Cleaning and inspection: The entire work area is cleaned in this step.
- Standardize: The places where things should be and the spaces where activities should be carried out are standardized.
- Discipline: Upon completion, feedback is provided to employees to make it a habit to keep work areas organized.

3.3.2.3 Step 3. Check

In this third stage, the efficiency of the previously implemented actions is verified. This is to ensure that the expected results are obtained. To this end, the new process capabilities are evaluated and the results obtained are monitored. We work with both systems (invoice and import), and if required, extra time is dedicated to the progress of the receipt of material. Once the receipt of the extra material has been completed, the invoice system is discontinued, and work is carried out for two weeks with the new systems only in order to standardize the process. Subsequently, we start measuring the new process flow, analyzing the new capabilities of the operations through observation, trial and error, and balancing the required personnel.

3.3.2.4 Step 4. Act

Once it has been verified that the actions are effective, in the fourth and final stage, the operating procedures are documented to standardize processes and operations. Subsequently, the necessary training is provided to the personnel involved, and metrics are established to monitor the process. Finally, the results achieved are reported to management.

3.4 Results

This section presents the results obtained by implementing the method described in the previous section. For a better understanding, these results are presented for each phase of the method described.

3.4.1 Findings from Phase 1: Plan

3.4.1.1 Flow Process Chart

Table 3.3 shows the initial flow process chart for the raw material receiving operation. As can be seen, 20 operations, 6 types of transport, 4 delays, 2 storage and 6 inspections were obtained. Observe that activity 5 includes an operation and an inspection simultaneously, so there are 37 activities. At first glance, it can be inferred that of these 37 activities, 17 of them are waste since they do not add value to the product and, therefore, generate costs for the company [14, 15]. However, more wastes may not be detected at first sight. Because of this, it is necessary to eliminate most of the activities that generate waste.

3.4.1.2 Ishikawa Diagram

Figure 3.10 illustrates the cause and effect diagram that resulted, listing the main problems for each of the six factors, such as machinery, method, labor, measurement, material and environments that affect the material receiving process taking up to 45 days.

The following is a detailed description of the causes detected for the problem analyzed for each of the six factors.

- *Machinery*: The scanner was inadequate; it did not read the supplier's code correctly, which caused the scanning operation to be repeated (often several times). In addition, the work tables were not suitable for the operation.
- *Method*: There was too much material handling and transfer, as the flow of operations was inefficient. Other causes were that there were too many unnecessary operations, the material took too long to arrive, there were several confirmations of the material received, the process was not standardized, and the change of color of labels in the labeling area was very frequent.
- *Manpower*: Personnel was not trained to operate and worked without clear goals or objectives on a day-to-day basis.
- *Measurement*: There was no control over imports received, so it was unknown when an import had been fully received. There were also no metrics in the process applied to the material.
- *Material*: The material had too many barcodes, and searching for the correct one was time-consuming. In addition, other materials lacked barcodes.
- *Environment*: No causes of the problem were detected.

Table 3.3 Flow process chart of raw material reception

No	Activity	●	➡	◗	▼	■	Description
1	Receipt of materials	●					Receipt and unloading of import materials
2	Barcode scanning	●					Barcode scanning of boxes and/or pallets
3	Confirmation of receipt					■	Confirmation of material received vs. Import/Export system
4	Delivery of documentation to the Traffic Department	●					Delivery of documents of material received to the Traffic Department
5	Confirmation of receipt	●				■	Confirmation of physical material vs. invoices received
6	Identification of pallets received	●					Strapping and identification of each pallet with its import number
7	Pallet movement to your assigned location		➡				Placement of pallets in their assigned locations
8	Pallet storage				▼		Storage of pallets until they are labeled
9	Collection of material to be labeled	●					Withdrawal of material to be labeled from the pallet
10	Confirmation of receipt					■	Comparison of physical material versus import document (invoice/packing list)
11	Placement of material in containers	●					Placement of material to be labeled in containers
12	Sending material to the labeling area		➡				Shipment of material to the labeling area with its respective invoice
13	Receipt of labeled material	●					The warehouse receives labeling material

(continued)

Table 3.3 (continued)

No	Activity	●	➡	◗	▼	■	Description
14	Supplier part number identification	●					Underline supplier part number
15	Barcode scanning or manual capture	●					Barcode or supplier part number scanning/capturing
16	Label printing	●					Label printing for scanned material
17	Material labeling	●					Place the label on the material
18	Log tag record	●					A logbook label will be printed and affixed to the invoice
19	Placement of rolls on trolleys	●					Placement of material on the trolley for transportation
20	Counting of labeled material by part number	●					Counting of labeled material by part number
21	Waiting for material to complete invoice			◗			Receipt of all invoice material must be completed
22	Transfer of material to incoming		➡				Transfer of labeling material to incoming for validation
23	Wait for incoming validation			◗			Waiting for validation by incoming
24	Validation of labeling by incoming					■	Incoming validation of labeling information
25	Material transfer to the warehouse		➡				Transfer of material from incoming to warehouse
26	Receipt of incoming material	●					The warehouse receives incoming material

(continued)

Table 3.3 (continued)

No	Activity	●	➡	◗	▼	■	Description
27	Wait for automatic validation			◗			Wait for automatic validation
28	Manual labeling validation					■	Automatic validation of labeling information
29	The capture of a photograph of the material	●					Taking of photographs for future traceability of labeled material
30	Placement of rolls on trolleys	●					Placement of material on the trolley for transportation
31	Transfers material to the front end area		➡				Moves the material cart to the front end area
32	Waiting for material confirmation			◗			Waiting for confirmation of material to be entered into the system
33	Confirmation of material received in the system					■	Counting of material received in the system vs. physical material
34	Transfers material to your location		➡				Transfer of material to your location
35	Gives entry to the material labeled in kárdex	●					KARDEX entry to the material received
36	Storage of materials				▼		Storage of materials at your location
37	Close completed invoices	●					Send an e-mail with complete invoices to close
Summary	Total						

(continued)

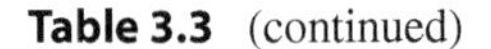

Table 3.3 (continued)

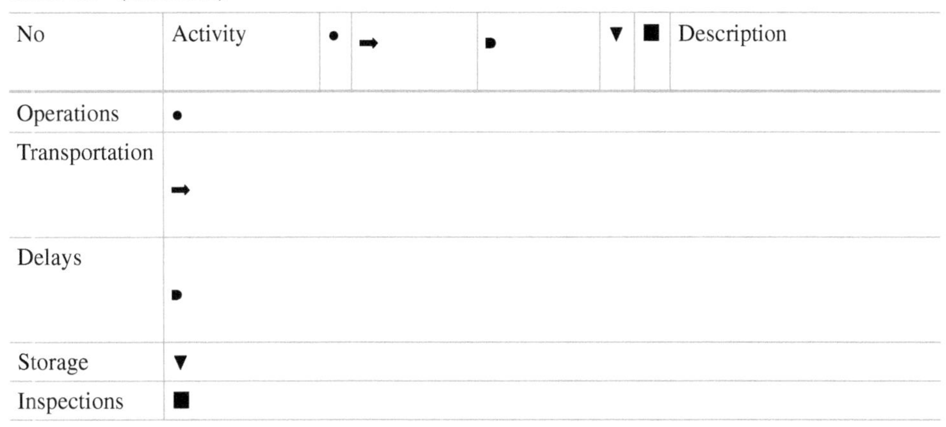

No	Activity	•	➡	◗	▼	■	Description
Operations	•						
Transportation	➡						
Delays	◗						
Storage	▼						
Inspections	■						

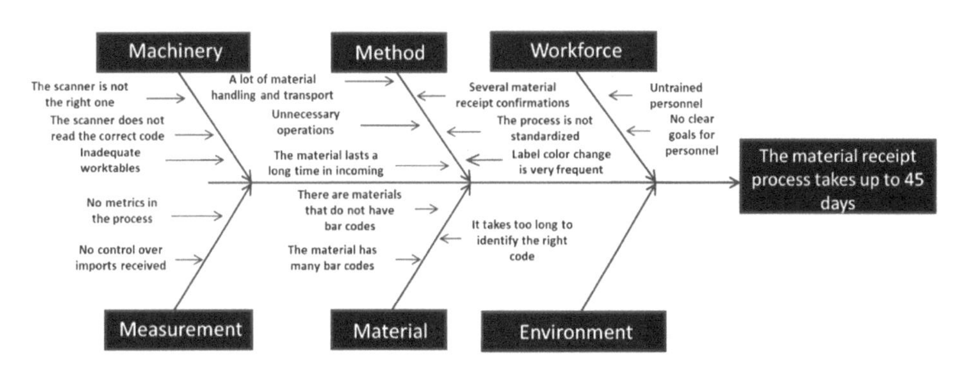

Fig. 3.10 Cause and effect diagram of the problem to solve

After analyzing the Ishikawa diagram, it can be seen that fifteen possible causes of the problem were detected and that the work method was the element where most of them were found, with a total of six. The actions defined to eliminate these causes and solve the problem are shown in Table 3.4.

3.4.2 Results of Step 2: Do

3.4.2.1 Standardizing the Label Color in the Labeling Area

It was decided to use the green label for all the products since it is the one that currently has more inventory, and to identify each of the customers; a different print design was used for each one. The system was able to recognize each customer and print the corresponding design.

Table 3.4 Planned activities for the solution of problems

Planned activity	Current status	Expected result
Standardization of the color of labeling labels	The company has 4 customers, and a different color is used for labeling materials for each of them (green, brown, yellow, and gray)	This action was intended to standardize and unify the color of the label, using the same label for all customers, eliminating the time required to change the printer label when labeling materials
Change in the materials receipt system from receipt by invoice to receipt by imports	Labeling the receipt of materials is done through the "CIMS–InvoicesWH" system. The person labeling receives the material along with the invoice and searches the system for the invoice and part number to be labeled. All part numbers included in the invoice must be completed. A logbook format printout is made on the invoice on which a logbook label is placed for each part number comprising the invoice. Once all the part numbers have been tagged, the closing of this invoice is requested to Planning by e-mail. The person can tag one invoice and then another without having finished the previous one, so the material of the first invoice is detained in the area, and until the material of this one is received and completed, it goes to the next operation	This action was intended to change the receipt system, where the import receipt is used instead of using the invoice for the receipt. The person labeling must complete the import receipt before proceeding to another one

(continued)

Table 3.4 (continued)

Planned activity	Current status	Expected result
Change of scanner in the labeling of materials	Each part number received is given a unique barcode label to track throughout the system's process. To do this, it is scanned after each process is completed. Here the worker has to search each roll of barcodes for the one that belongs to the supplier's part number, so it takes too much time to search and makes it easy to make mistakes Scanning can be automatic or manual	This action is intended to install a new barcode scanner that can scan all 15 barcodes on the material simultaneously, reducing the time spent searching for the correct supplier information and eliminating possible errors. In addition, it is also intended to introduce an optical recognition system to read the printed characters without having to type them
Analysis of operations and elimination of changes in the process method		With this analysis, the plan is to identify and eliminate the changes in the method, starting from the flow process chart to evaluate each of the operations and determine whether or not it is an operation that adds value to the process. The following questions help in this decision: Can the operation be eliminated? Can it be combined with another? Can it be done in the idle time of another? Is the sequence of activities the best possible? Should the operation be performed in another department or location to save cost and handling?
Application of 5's in the process	Unorganized, dirty and unsafe workstations. There is no standardization of workstations, and the spaces for performing activities are not delimited	This action aims to create more organized, orderly, clean and safe workplaces that allow workers to work more productively, improving their working environment

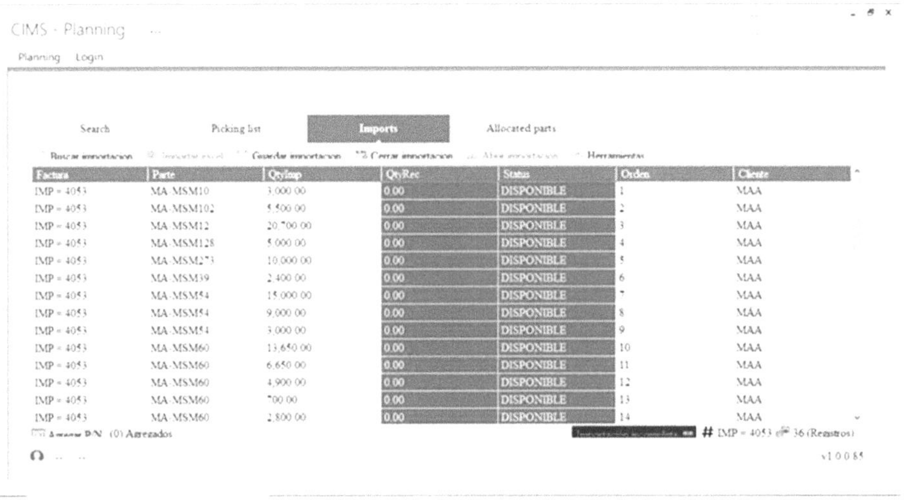

CIMS - Planning

Planning Login

Search | Picking list | Imports | Allocated parts

Buscar importacion | Guardar importacion | Cerrar importacion | Herramientas

Factura	Parte	QtyImp	QtyRec	Status	Orden	Cliente
IMP = 4053	MA-MSM10	3,000.00	0.00	DISPONIBLE	1	MAA
IMP = 4053	MA-MSM102	5,500.00	0.00	DISPONIBLE	2	MAA
IMP = 4053	MA-MSM12	20,700.00	0.00	DISPONIBLE	3	MAA
IMP = 4053	MA-MSM128	5,000.00	0.00	DISPONIBLE	4	MAA
IMP = 4053	MA-MSM273	10,000.00	0.00	DISPONIBLE	5	MAA
IMP = 4053	MA-MSM39	2,400.00	0.00	DISPONIBLE	6	MAA
IMP = 4053	MA-MSM54	15,000.00	0.00	DISPONIBLE	7	MAA
IMP = 4053	MA-MSM54	9,000.00	0.00	DISPONIBLE	8	MAA
IMP = 4053	MA-MSM54	3,000.00	0.00	DISPONIBLE	9	MAA
IMP = 4053	MA-MSM60	13,650.00	0.00	DISPONIBLE	10	MAA
IMP = 4053	MA-MSM60	6,650.00	0.00	DISPONIBLE	11	MAA
IMP = 4053	MA-MSM60	4,900.00	0.00	DISPONIBLE	12	MAA
IMP = 4053	MA-MSM60	700.00	0.00	DISPONIBLE	13	MAA
IMP = 4053	MA-MSM60	2,800.00	0.00	DISPONIBLE	14	MAA

(0) Agregados # IMP = 4053 36 (Registros)

v1.0.0.85

Fig. 3.11 Cims system–planning

3.4.2.2 Change in the Materials Receipt System from Receipt by Invoice to Receipt by Imports

A new system called "Cims–Planning" was developed. This system changes the initial method of receipt of materials, which was an invoice, to the method of receipt of materials by import. Figure 3.11 shows this system.

This system allowed to control of receipts of imports in an orderly manner since it was required to receive all the import material in a single exhibition before moving on to the next. In addition, the supplier's part number was automatically recognized by scanning the barcode with the vendor's information, making the automatic relationship with the internal part number, and eliminating the search and selection of the number by the labeler. Once the material was validated, the system automatically discounted the material, and the system closed when all the material was discounted from the import.

3.4.2.3 Changing the Scanner in the Labeling Area

The traditional trigger scanner was replaced for automatic labeling, and a new Keyence SR-2000W barcode scanner was installed (Fig. 3.12), which allows scanning all 15 barcodes on the reel in one shot or simultaneously, including taking a photo of the material. With this scanner, it was not necessary to find the supplier's barcode; the material was placed under the scanner, and the system scanned all the barcodes at once and recognized them automatically.

The system was adapted for this scanner since the person only selects the import to work with and scans the material, while the system worked by itself.

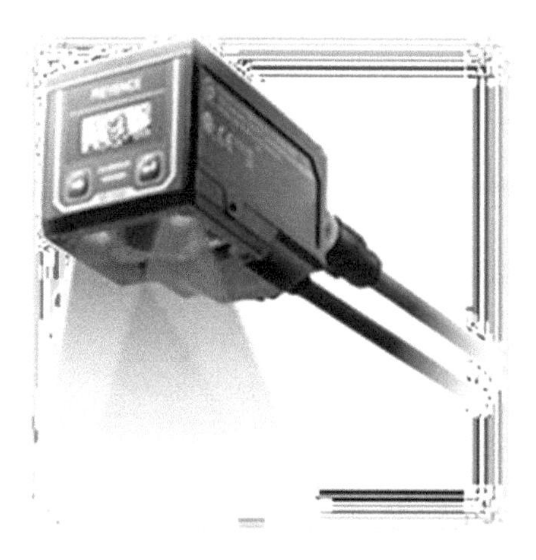

Fig. 3.12 Keyence SR-2000W scanner

Fig. 3.13 Keyence CA-500CX vision camera

The OCR (Optical character recognition) system was implemented for manual scanning, using a Keyence CA-H500CX vision camera, as illustrated in Fig. 3.13.

The system recognized the part number, and the label was printed, eliminating the inspection by the Incoming Quality Control (IQC) area. By automating the operation, the person only had to select the import to work with and the material supplier to perform the library search.

3.4.2.4 Operation Analysis and Waste Elimination in the Process Method

Table 3.5 shows the conclusions obtained for each raw material receipt process operation.

With this, 9 operations, 1 storage, 2 types of transport, 4 delays and 4 inspections were eliminated, leaving only 11 operations, 4 types of transport, 1 storage, 3 inspections and zero delays, as shown in Table 3.6 of the new flow process chart. In the end, only 12 operations, 4 types of transport, 1 storage, 3 inspections, and no delays remained, leaving only 20 activities.

Table 3.5 Results of operations analysis

No	Activity	●	➡	D	▼	■	Is it necessary?	Waste it generates
1	Receipt of materials	●					Yes	
2	Barcode scanning	●					Yes	
3	Confirmation of receipt					■	No	Overprocessing
4	Delivery of documentation to the Traffic Department	●					Yes	
5	Confirmation of receipt					■	No	Overprocessing
6	Identification of pallets received	●					Yes	
7	Movement of pallets to your assigned location		➡				Yes	Transportation
8	Pallet storage				▼		No	Inventory
9	Collection of material to be labeled	●					Yes	
10	Confirmation of receipt					■	No	Overprocessing
11	Placement of material in containers	●					Yes	
12	Sending material to the labeling area		➡				Yes	Transportation
13	Receipt of labeled material	●					Yes	
14	Supplier part number identification	●					No	Overprocessing
15	Barcode scanning or manual capture	●					No	Overprocessing
16	Label printing	●					No	Overprocessing
17	Material labeling	●					Yes	
18	Log tag record	●					No	
19	Placement of rolls on trolleys	●					Yes	

(continued)

Table 3.5 (continued)

No	Activity	●	➡	◗	▼	■	Is it necessary?	Waste it generates
20	Counting of labeled material by part number	●					No	Overprocessing
21	Waiting for material to complete invoice			◗			No	Waiting time, stock
22	Transfer of material to incoming		➡				No	Transportation, unnecessary movement
23	Wait for incoming validation			◗			No	Waiting time, stock
24	Validation of labeling by incoming					■	No	Overprocessing
25	Material transfer to the warehouse		➡				No	Transportation, unnecessary movement
26	Receipt of incoming material	●					No	Overprocessing
27	Wait for automatic validation			◗			No	Waiting time, stock
28	Manual labeling validation					■	Yes	Overprocessing
29	The capture of photograph of the material	●					Yes	
30	Placement of rolls on trolleys	●					Yes	
31	Transfer of material to the front end area		➡				Yes	Transportation
32	Waiting for material confirmation			◗			No	Waiting time
33	Confirmation of material received in the system					■	No	Overprocessing
34	Material transfer to your location		➡				Yes	Transportation

(continued)

Table 3.5 (continued)

No	Activity	●	➡	◗	▼	■	Is it necessary?	Waste it generates
35	Entry of labeled material in kárdex	●					Yes	
36	Storage of materials				▼		Yes	Stock
37	Closing of completed invoices	●					No	Overprocessing

3.4.2.5 Application of 5's in the Process

Previously, there was no delimitation for material flow or standardization in the workstations, which caused errors in the activities. As a result of the application of the 5's, delimited spaces for carrying out activities, material and personnel flow, and standardized workstations were obtained. In addition, more organized, orderly, clean and safe workplaces were obtained, allowing workers to work more productively, and improving the work environment, as occurred in the case study of [16]. Figure 3.14 shows the standardized labeling area.

3.4.3 Results of Step 3: Check

3.4.3.1 Implementation and Results

After having implemented the actions of the previous stage, it took about 5 weeks to complete the receipt of the material pending invoices. Table 3.7 shows the results of implementing the actions proposed in step 1. As can be seen, the capacity of the discarding operation increased from 1,820 units/day to 4,610 units/day (an increase of 253.03%), and that of the labeling operation increased from 580 units/day to 1,470 units/day (an increase of 253.47%), while the storage capacity increased from 1,720 units/day to 4,520 units/day (an increase of 262.79%). Finally, the total number of workers required decreased from 14 to 11.

During September 2021, imports behaved as shown in Fig. 3.15. As can be seen, it took a maximum of 7 days to close two imports and less than one day to close two others. Most of them were closed in a single day.

As for overtime, it was no longer required for the material receipt process as of September since, with the results obtained in stage 2, the material flowed more quickly. Figure 3.16 shows the monthly overtime trend for the material receipt process during 2021.

Table 3.6 New flow process chart for receipt of raw material

No	Activity	■	➡	◗	▼	♦	Description of the operation
1	Receipt of materials	■					Receipt and unloading of import materials
2	Barcode scanning	■					Barcode scanning of boxes and/or pallets
3	Confirmation of receipt					♦	Confirmation of material received vs. Import/Export system
4	Delivery of documentation to the Traffic Department	■					Delivery of documents of material received to the Traffic Department
5	Identification of pallets received	■					Strapping and identification of each pallet with its import number
6	Movement of pallets to your assigned location		➡				Placement of pallets in their assigned locations
7	Collection of material to be labeled	■					Withdrawal of material to be labeled from the pallet
8	Confirmation of receipt					♦	Comparison of physical material vs. import document (invoice/packing list)
9	Placement of material in containers	■					Placement of material to be labeled in containers
10	Sending material to the labeling area		➡				Shipment of material to the labeling area with its respective invoice
11	Receipt of labeled material	■					The warehouse receives material for labeling
12	Scanning of supplier information	■					Supplier information scanning and label printing
13	Material labeling	■					Place the label on the material
14	Automatic label validation					♦	Automatic validation of labeling information and automatic photo capture
15	Placement of rolls on trolleys	■					Placement of material on the trolley for transportation

(continued)

Table 3.6 (continued)

No	Activity	■	➡	●	▼	♦	Description of the operation
16	Transfer of material to the front end area		➡				Moves the material cart to the front end area
17	Transfer of material to your location		➡				Transfer of material to your location
18	Entry of labeled material in kárdex	■					KARDEX entry to the material received
19	Storage of materials				▼		Storage of materials at your location

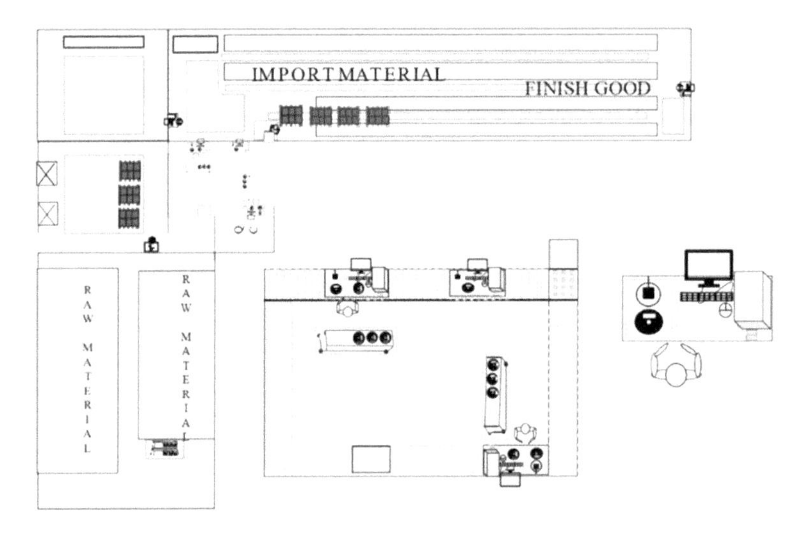

Fig. 3.14 Standardized warehouse area

Table 3.7 New capacity of operations (per day)

Operation	Capacity (units)	Total number of persons per operation (3 shifts)
Receipt of raw material		1
Discarded	4,610	
Tagged	1,470	
On-site warehousing	4,520	
Total	10,600	

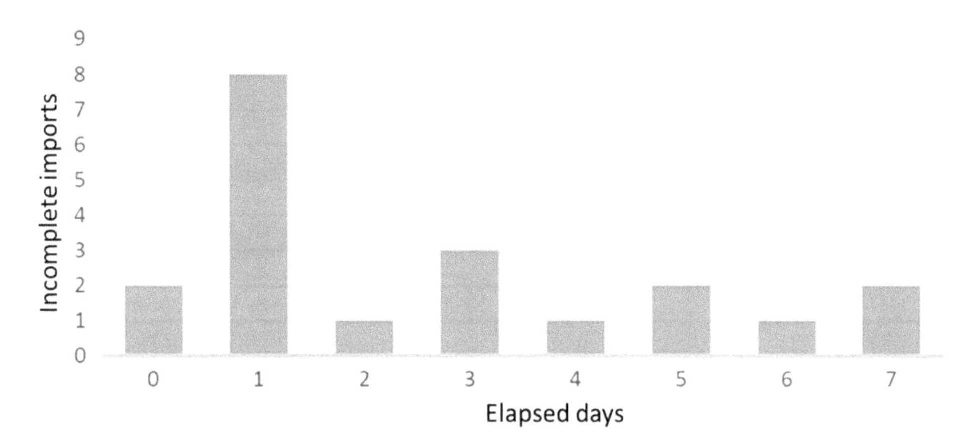

Fig. 3.15 Elapsed time to complete imports

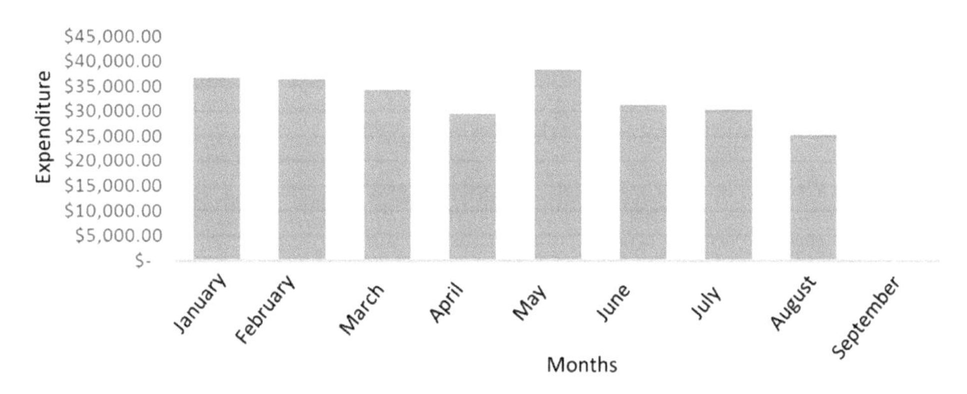

Fig. 3.16 Monthly trend of overtime during 2021

3.4.4 Results of Step 4: Act

At this stage, instruction sheets were obtained for the different operations of the raw material receipt process. Figure 3.17 shows the result of the instruction sheet for the material labelling and validation operation. Similar results were obtained for the other operations. With this, the results mentioned above were kept stable to date (July 2022), avoiding the recurrence of the initial problems. Figure 3.18 shows the resulting flow chart for the material receipt process.

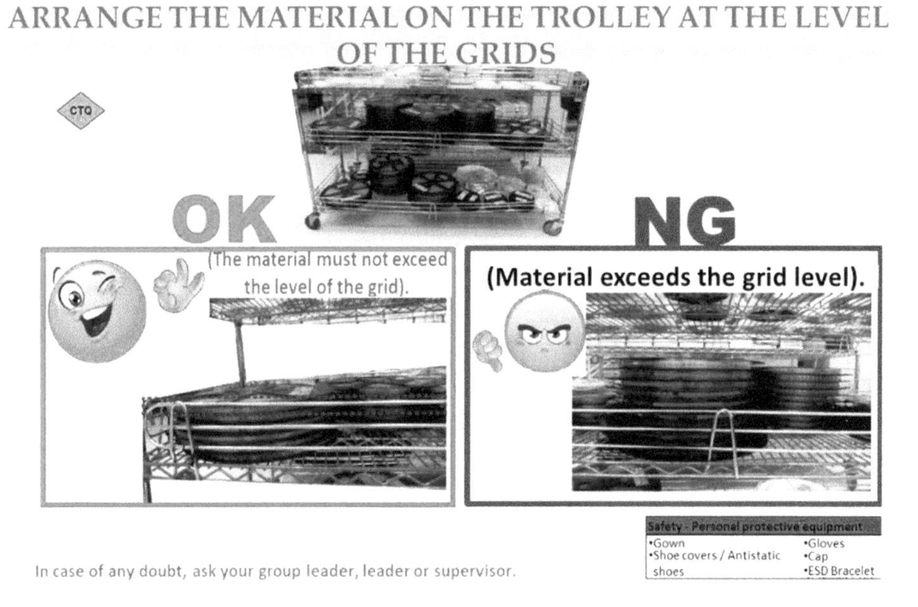

Fig. 3.17 Instruction sheet for trolley roll placement operation

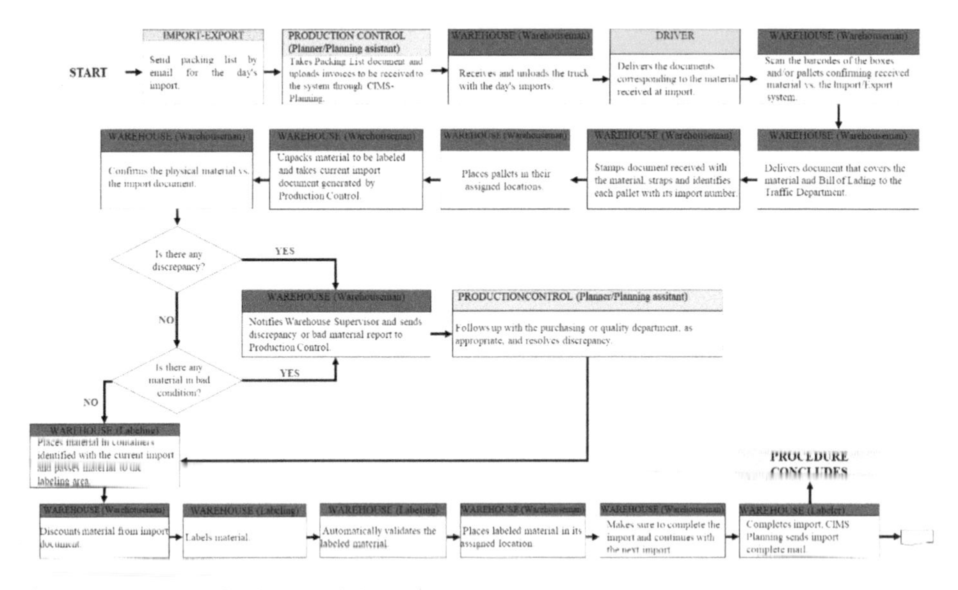

Fig. 3.18 Flowchart for the materials receipt process

3.5 Conclusions

The project's overall objective was to reduce the time of the materials receiving process. Regarding this objective, it is concluded that it was achieved since, initially, the process took up to 45 days, and the results indicated that the process currently takes a maximum of 7 days, which implies a reduction of 84.44%. This result agrees with that obtained by Lerche et al. [17], who applied the PDCA cycle in a case study in the construction industry, obtaining a 28% reduction in the delivery time of materials. Similar results were obtained by Widodo and Fardiansyah [18] and Radhila [19], who were able to reduce the delivery process time in a warehouse by 77% and the idle time by 78%, respectively, by applying the PDCA cycle.

Regarding the specific objectives, the first one was to increase the labeling operation by 50% of the capacity of processed units. Initially, the capacity of this operation was 580 units/shift, and after the changes were made, the capacity was 1470 units/shift, increasing by 253%, so it can be concluded that this objective was met. This result is similar to that obtained by Milosevic et al. [20], who applied the PDCA cycle in a company to produce welded excavator frames, which occurred due to the increase in customer demand derived from the variation of the global market. As a result, the processing capacity increased by 10.67%.

The second objective was to reduce 20% of the raw material receiving process operations. Initially, the process consisted of 20 operations. Once the project was completed, the total number of operations was 12, reducing operations by 40%, concluding that this objective was met. In addition, 1 storage, 2 types of transport, 4 delays and 4 inspections were eliminated. This result is consistent with what was mentioned by Mutingi et al. [21], who indicate that lean manufacturing tools, including the PDCA cycle, are useful for identifying and eliminating (or reducing) waste, defined in terms of transportation, over-processing, inventory (storage), and waiting (delays).

Finally, the third objective was to eliminate 100% of the extra time required for the raw material receipt process. The results indicate that once the accumulated material in the process was cleaned, it was no longer necessary to invest extra time to work on the receipt of materials; therefore, it is concluded that this objective was also achieved. This result is consistent with Anoye and Adama [22], who applied PDCA to eliminate overtime in a bar soap manufacturing company.

Generally, the PDCA cycle, supported by tools such as the flow chart, the cause-effect diagram, and the 5's, helps reduce process times, increase process capacity, and eliminate unnecessary operations and extra work time.

References

1. Schwab K, Sala-i-Martín X (2013) The global competitiveness report 2013–2014. In: Schwab K (ed) World Economic Forum, Geneva, 551 pp
2. Eisen K, Eifert T, Herwig C, Maiwald M (2020) Current and future requirements to industrial analytical infrastructure—part 1: process analytical laboratories. Anal Bioanal Chem 412:2027–2035
3. Kolinski A, Sliwczynski B (2015) Evaluation problem and assessment method of warehouse process efficiency. In: Proceedings of the 15th international scientific conference business logistics in modern management [Internet]. Osijek, Croatia [cited 2022 May 18], pp 175–188. https://hrcak.srce.hr/ojs/index.php/plusm/article/view/3880
4. Tönnissen S, Teuteberg F (2018) Using blockchain technology for business processes in purchasing—concept and case study-based evidence. In: Abramowicz W, Paschke A (eds) Business information systems. Springer, Berlin, Germany, pp 253–264
5. Moch R, Riedel R, Müller E (2014) Key success factors for production network coordination. In: Zaeh M (ed) Enabling manufacturing competitiveness and economic sustainability proceedings of the 5th international conference on changeable, agile, reconfigurable and virtual production. Springer, Cham, Munich, Germany, pp 327–332
6. Costa F, Carvalho M do S, Fernandes JM, Alves AC, Silva P (2017) Improving visibility using RFID—the case of a company in the automotive sector. Procedia Manuf 13:1261–1268
7. Sarrafha K, Rahmati SHA, Niaki STA, Zaretalab A (2015) A bi-objective integrated procurement, production, and distribution problem of a multi-echelon supply chain network design: a new tuned MOEA. Comput Oper Res 1(54):35–51
8. Mahendrawathi E, Astuti HM, Wardhani IRK (2015) Material movement analysis for warehouse business process improvement with process mining: a case study. In: Bae J, Suriadi S, Wen L (eds) Asia Pacific business process management third Asia Pacific conference. Springer Cham, Busan, South Korea, pp 115–127
9. Bradford M, Gerard GJ (2015) Using process mapping to reveal process redesign opportunities during ERP planning. J Emerg Technol Account 12(1):169–188
10. Fauzan R, Shiddiq MF, Raddlya NR (2020) The designing of warehouse management information system. In: IOP conference series: materials science and engineering. IOP Publishing, Bandung, Indonesia, pp 1–7
11. Wutthisirisart P, Sir MY, Noble JS (2015) The two-warehouse material location selection problem. Int J Prod Econ 1(170):780–789
12. Frontoni E, Marinelli F, Rosetti R, Zingaretti P (2020) Optimal stock control and procurement by reusing of obsolescences in manufacturing. Comput Ind Eng 1(148):106697
13. Cortinhal MJ, Lopes MJ, Melo MT (2019) A multi-stage supply chain network design problem with in-house production and partial product outsourcing. Appl Math Model [Internet] [cited 2022 May 18] 70:572–594. https://www.sciencedirect.com/science/article/pii/S0307904X19300769
14. Shah D, Patel P (2018) Productivity improvement by implementing lean manufacturing tools in manufacturing industry. Int Res J Eng Technol [Internet] [cited 2018 Oct 3] 5(3):3794–3798. www.irjet.net
15. Wahab ANA, Mukhtar M, Sulaiman R (2013) A conceptual model of lean manufacturing dimensions. Procedia Technol 1(11):1292–1298
16. Srinivasan S, Ikuma LH, Shakouri M, Nahmens I, Harvey C (2016) 5S impact on safety climate of manufacturing workers. J Manuf Technol Manag 27(3):364–378

17. Lerche J, Neve H, Wandahl S, Gross A (2020) Continuous improvements at operator level. J Eng Proj Prod Manag 10(1):64–70
18. Widodo T, Fardiansyah I (2019) Implementasi continuous improvement Dengan Menggunakan Metode Pdca Pada Proses Handover di Warehouse PT. ABC. J Ind Manuf [Internet] 4(1):37–44. http://jurnal.umt.ac.id/index.php/jim/article/view/1243
19. Radhila A (2018) Implementasi warehouse management Menggunakan Metode PDCA Studi Kasus Di CV. Innotech Solution -Malan. J Valtech [Internet] 1(1):230–241. https://ejournal.itn.ac.id/index.php/valtech/article/view/219
20. Milosevic M, Djapan M, D'Amato R, Ungureanu N, Ruggiero A (2021) Sustainability of the production process by applying lean manufacturing through the PDCA cycle—a case study in the machinery industry. In: Hloch S, Klichová D, Pude F, Krolczyk GM, Chattopadhyaya S (eds) Advances in manufacturing engineering and materials II [Internet]. Springer, Cham, Nový Smokovec, Slovakia [cited 2022 Jul 4], pp 199–211. https://link.springer.com/chapter/10.1007/978-3-030-71956-2_16
21. Mutingi M, Isack HD, Kandjeke H, Mbohwa C (2017) Adoption of lean tools in medical laboratory industry: a case study of Namibia. In: Proceedings of the 2017 international conference on industrial engineering and operations management (IEOM). IEOM Society International, Bristol, UK, pp 895–901
22. Anoye BA, Adama O (2016) Continual improvement in small soaps company. Ethics Crit Think J [Internet] [cited 2022 May 26] 2016(1):6–75. https://web.p.ebscohost.com/abstract?direct=true&profile=ehost&scope=site&authtype=crawler&jrnl=15475425&asa=Y&AN=118959543&h=N%2BjPKzlaoxiDUTozBsvoT21CVLoL8PsW6k9QhO2pVyUQJO4O4Y%2FJrceWrPycOdM1TRBCsYRauCjg5C7jehJzxg%3D%3D&crl=c&resultNs=AdminWebAuth&resu

4 Case Study 3. Eliminating Waste and Increasing Performance

4.1 Introduction

Lean manufacturing focuses on eliminating company waste, which is anything that does not add value to a product [1]. According to the literature, there are seven wastes in manufacturing: overproduction, defects, inventory, transportation, waiting time (delays), movement, and overprocessing [2]. In the case of overproduction, this waste arises from the production of a greater number of goods in the workplace to avoid problems caused by a possible failure of a machine or the production completed before the delivery time [3]. According to several authors, this waste is the most crucial and generates many problems, other waste and costs for companies [3, 4].

Regarding defect waste, the literature indicates that defects in internal failures refer to scrap, rework, and delay, while external failures include warranty, repairs, and field service [4]. In addition to physical defects, which directly increase the cost of goods sold, there can also be errors in paperwork, delayed delivery and production, according to incorrect specifications, as well as the use of too many raw materials or the generation of unnecessary waste [2, 5].

Wahab et al. [4] mention that defects are both short-term and long-term direct costs, and defects in Toyota Production System are an opportunity to improve rather than negotiate. When a defect occurs, rework may be necessary; otherwise, the product will be discarded. The generation of defects will not only waste material and labor but also create material shortages, make it difficult to meet schedules, create downtime at downstream work centers, and extend the lead time for manufacturing [2, 6].

There are three types of inventory waste: raw material, work in process, and finished goods [4, 7]. Excess inventory leads to higher inventory financing costs, higher warehousing costs, higher defect rates, tends to increase lead time, prevents quick identification of problems, and increases space requirements, which affects communication [4, 7]. To carry

A. Realyvásquez Vargas et al., *The PDCA Cycle for Industrial Improvement*, Synthesis Lectures on Engineering, Science, and Technology,
https://doi.org/10.1007/978-3-031-26805-2_4

out effective purchasing, it is especially necessary to eliminate inventory due to incorrect delivery times [7].

Transportation waste includes the movement of materials that do not add value to the product, such as moving materials between workstations [7], as well as double handling [4]. Unnecessary material movements can cause damage and deterioration in materials with the communication distance between processes, which affects the productivity and quality of the products [4], and the literature indicates that this waste is caused by poor organization in the workplace [8]. Transportation of materials between different processing steps prolongs production cycle times and inefficient use of labor and space [7]. Any transportation of materials within companies can be considered waste.

Concerning delay waste, this is defined as downtime for workers or machines due to bottlenecks or inefficient production flow on the shop floor [7]. According to Shah and Patel [7], this waste occurs when time is used inefficiently and includes delays between processing units because they are not moving or not being worked on. These authors mention that this wastage affects both goods and workers, each of whom spends time waiting that can be used for training or maintenance activities and should not result in overproduction [7]. Finally, this waste tends to increase the lead times of products, whether in process or finished [4].

Movement wastage refers to any unnecessary physical movement or walking by workers that diverts them from the work or activity they are performing [7]. This could include walking around the factory to fetch a tool or even unnecessary or difficult physical movements due to a lack of ergonomics, which slows workers down. This waste encompasses human and design dimensions [7]. The human dimensions relate to the ergonomics of production, where operators have to stretch, bend, pick up, or move around to get a better view [2, 6]. Such waste is tiring for employees and will likely lead to productivity and quality problems [4]. The dimensions of the physical plant layout refer to a poorly designed workplace [4].

Finally, overprocessing waste refers to unintentionally performing more processing work than is required by the customer in terms of product quality or features, such as polishing or applying finishes to some areas of the product that will not be seen by the customer [7]. Overprocessing occurs when overly complex solutions are found to simple procedures, such as using a large machine o to process a product [4, 7], an inflexible machine instead of several small and flexible ones.

It also refers to machines and processes that do not have quality and capability. Wahab et al. [4] declare that a capable process requires correct methods, training and standard that does not result in defects. These authors also mention that over-complexity often discourages ownership and encourages employees to overproduce to recoup the large investment in complex machines. This approach encourages poor distribution, leading to excessive transportation and poor communication. The ideal is to have the smallest possible machine, capable of producing the required quality, located next to the upstream and downstream operations [4].

Therefore, it is concluded that waste represents costs for companies and is related to each other. However, there are successful case studies reported in the literature where lean manufacturing tools have been applied to reduce or eliminate these wastes. One of the most commonly used tools is the Plan-Do-Check-Act cycle (PDCA), and some of these case studies are mentioned below.

Jangid [9] implemented the PDCA cycle in a manufacturing company, eliminating overproduction. For their part, Ikatrianasari and Putra [10] applied PDCA in a multinational company in the connector and terminal industry and reduced the transportation time of materials and products by 70.96 h (lead time); products in the selection buffer stock (inventory) were eliminated, and the percentage of defective products decrease by 82.6% (defects). In addition, 51 product inspection activities (over-processing) were eliminated, and the addition of a production monitor resulted in an increase in productivity from 86.5 to 91.7%.

Similarly, Nurafiqah [11] applied the PDCA cycle in a process improvement project in a bakery in Malaysia. Findings in this project were a reduction in the search time (delay) of equipment and materials from 10 min with 4 s to 5 min with 25 s, representing an improvement of 53.81%.

In another research, Nino et al. [12] implemented the PDCA cycle in a Sterile Processing Department (SPD) of a Rural Hospital in which there was wasted waiting time in surgeries that were delayed because one or more trays were not available, as well as trays ready to be sterilized, waiting because the steam sterilizer was busy. This problem was magnified when the steam sterilizer was used to process only one tray. In addition, SPD technicians were waiting for the availability of a computer. As a result of implementing the PDCA cycle, the second computer became available for use, reducing the waiting time for technicians. In addition, given the new process organization, the SPD staff identified which cart was being filled to be taken to the steam sterilizer and which cart had already finished the sterilization process. The process flow became more stable, and unsterilized trays were no longer sent to the surgery room by mistake and trays were not sterilized twice.

These case studies demonstrate that correctly applying the PDCA cycle can significantly impact companies. The following is a case study where the PDCA cycle is applied to redesign a cell and increase production capacity in a manufacturing company.

4.2 Case Study

The case study presented in this chapter takes place in a company that is a leading provider of interconnect and transmission solutions, serving the telecommunications, data transmission, industrial and medical sectors. This company is a leading producer of interconnecting cable assemblies containing optical fiber. In addition, it manufactures high-speed copper, radio frequency, standard industrial cable assemblies, and a wide range of power cables.

The company operates eight manufacturing plants worldwide, with extensive product development, manufacturing and testing capabilities. It also has more than twenty sales and support offices strategically located in Asia, America and Europe. This case study takes place in Tijuana, Mexico, a plant dedicated to developing and marketing assemblies for different industries such as electrical, automotive, electronics, medical, aerospace, telecommunications, plugs, harnesses, MRI cables, and power transmission, among others.

The company's organizational structure has several departments, which allow achieving a final product with quality, as required by the customer. For this project, New Product Introduction and Manufacturing areas will be involved. Figure 4.1 shows the organizational chart of the company.

The following briefly describes the company's process flow, starting with the customers (stakeholders) and then with each department (see Fig. 4.2).

Stakeholders: The manufacturing process of an assembly starts from the need of a customer, shareholder or community to use the company's services to assemble a product. When there is a need, 4 departments are involved; on the one hand, the risk management process is in charge of evaluating and managing the activities before executing them; the resource management process is in charge of making sure that all the material is ready to be used for manufacturing and it is here when the result of both operations is evaluated in an environmental management system to determine that the product to be manufactured is within the standards. On the other hand, the DCU (Document Control

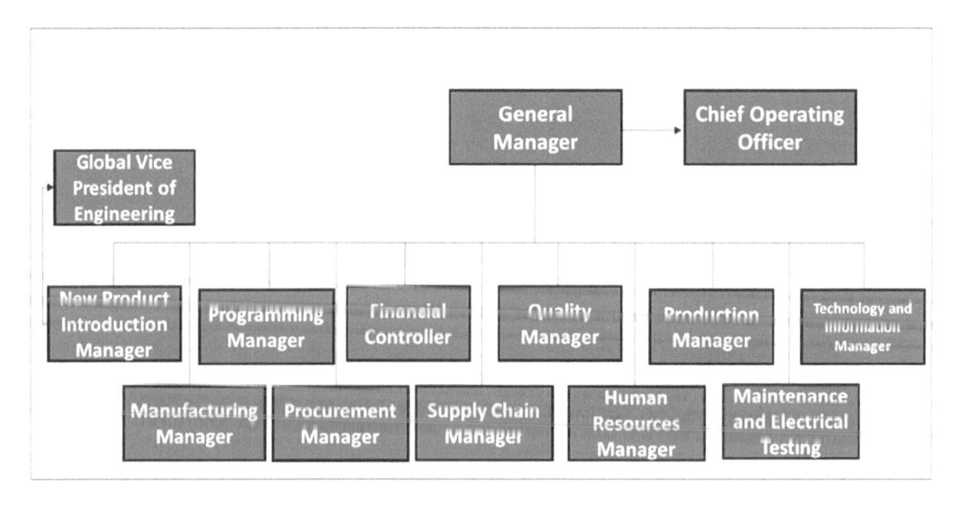

Fig. 4.1 Organisational chart of the company

Fig. 4.2 Process map of the company

Unit) department is in charge of releasing the documentation in conjunction with the management responsibility process to proceed with the product's manufacture.

Planning, NPI (New Product Introduction) and the Purchasing department: In the first phase, the Planning and NPI departments agree with Purchasing to establish a date for purchasing materials needed to build the product. In this way, the Planning Department is in charge of setting the shipping dates.

Purchasing, Warehouse and Quality: Once the Purchasing department acquires everything necessary for assembling the product, the raw material is taken to the Warehouse Department, which is classified and reviewed by the Quality Department. This department is in charge of evaluating the materials and confirming that they are in full use to proceed with the manufacture or assembly of the final product.

NPI (New Product Introduction) or Production: Once the material is released and in full use for construction, it can be of two types: (1) confirmed first article, and (2) sample. In both cases, the NPI department is in charge of manufacturing or assembly. In this department, the first visual aids and primary documentation are made, such as a control plan and process mode and failure, so the sample or first article being introduced becomes mass production.

Once converted to mass production, the first item or sample is passed to the Production Department. This department is responsible for standardizing the production process and making improvements to make it more robust. Once the part numbers are finished as a final product, they are sent to the Warehouse Department to an area called "Shipping", which is in charge of getting the final product to the customer.

Figure 4.2 shows the process map of the company, which involves the departments mentioned above. On the other hand, Fig. 4.3 shows the distribution of these departments throughout the plant.

The company has approximately 750 employees and two general shifts, the morning shift from 7:00 a.m. to 5:30 p.m. from Monday to Thursday and the output on Fridays at 3:00 p.m. The second shift is the night shift which runs from 7:00 p.m. to 5:30 a.m. from Monday through Thursday mornings. There is a third shift exclusively for production personnel of a specific client, from 7:00 a.m. to 7:00 p.m., from Monday to Thursday. This personnel works in the area known as JCI, where the study and analysis were carried out to increase the capacity of the production cell 2311, focusing on one model only, the part number called XYZ.

Figure 4.4 shows the location of cell 2311 within the area. It is in this cell where the high voltage cables are manufactured, and its manufacturing process has several areas of opportunity for improvement, and this causes low production, in addition to using resources that are not necessary.

The product is manufactured as follows:

1. The wire is taken from the harness and identified according to its length and gauge, i.e., its designator.
2. The cable is passed to the second station, where it is semi-stripped with the automatic stripper, and, finally, it is manually stripped with a knife.

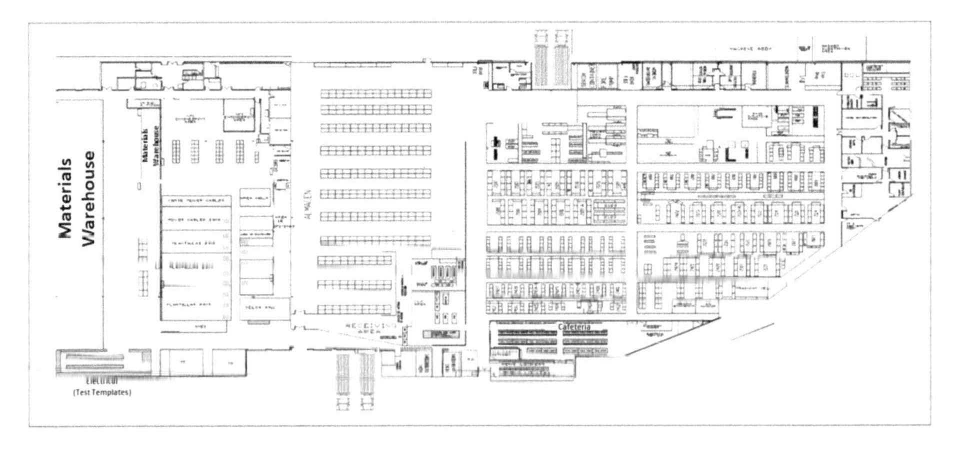

Fig. 4.3 Layout of the company

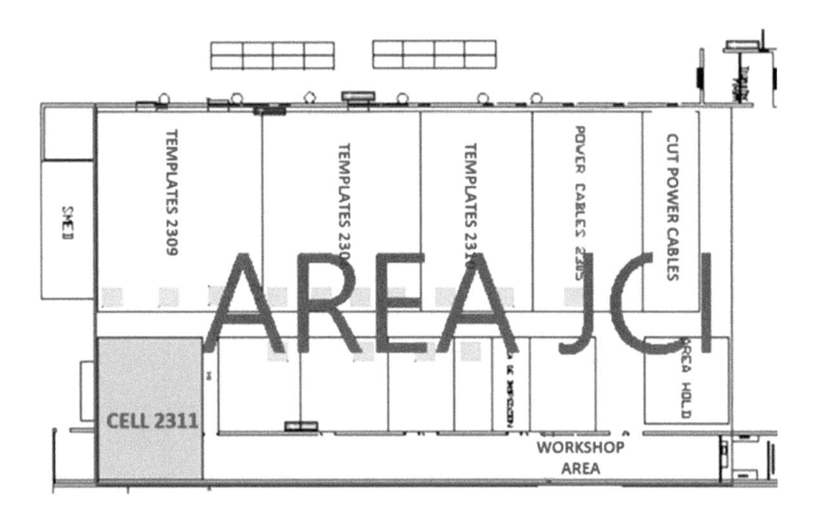

Fig. 4.4 JCI area layout

3. Once stripped, the cable is riveted according to its designator. The drawing and visual aids are observed to identify the orientation of the terminals and which terminal (part number) goes according to the designator.
4. Once riveted, it is passed to the inspection operation, where it is verified that the position of the terminals and the length is correct according to the designator indicated on the tape. Once the part is validated, the tape is removed, and the designator labels are placed.
5. Finally, the part is passed to the quality inspector for approval and packaging.

Some operations are not required within the process, such as labeling the cable twice. These aspects are the ones that mainly need to be improved to increase the cell's production; the lack of a material rack to place and separate the material causes disorder and accumulation of raw material, generating problems in the cell. On the other hand, the area's distribution is inadequate since relatively long transports are generated to obtain the raw material, leading to another waste: the waiting time. Currently, the cell has a cycle time of approximately 1:30 to 2:00 min per operation, and it is expected to reach 53 s of cycle time to increase the productivity of the cell.

To perform the operations, cell 2311 has a semiautomatic stripper, which performs the function of marking the cable for stripping. This stripper is obsolete and does not perform its function 100% since it has been used for several years and has lost its functionality. On the other hand, for the labeling operation, templates are used for each of the models, i.e., even though the lengths of the cables and the labeling are similar or the same, a change of template must be made for each model, making up to 3 model changes per day. Figure 4.5 shows the current layout of the assembly cell.

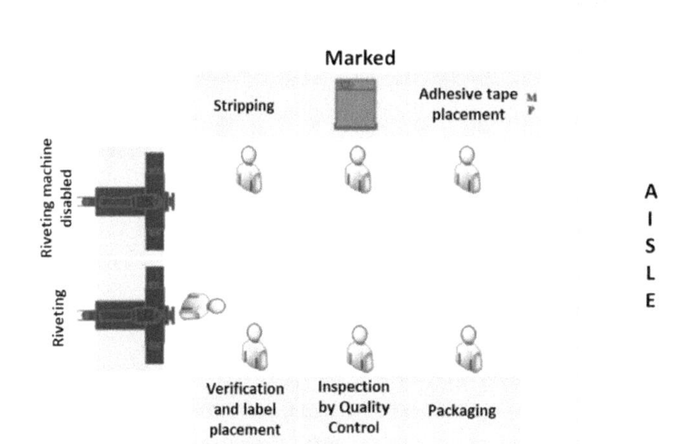

Fig. 4.5 Layout of cell 2311

4.2.1 Problem Statement

The company is dedicated to the manufacture of cables and harnesses. It has a cell dedicated to manufacturing high voltage cables, which is the product that currently generates more money income, as it is expensive and has constant demand and it is the most valuable product for the company; however, the manufacturing process is not the optimal one to satisfy the demand from the client and to fulfill the quality specifications at the first time. That problem is because, during the process, there is waste such as over-processing, unnecessary movements, waiting times, and defects, among other wastes that do not add value to the product.

As a result, the desired production rate is not being achieved. One of the main defects in high voltage cables is the misidentification in pieces; that is to say, their designation does not coincide with the length and gauge specifications shown in the drawing. Also, within the process, unnecessary operations are identified since the process is not correctly defined, which causes a cycle time longer than what is desired in order to meet the demand required by the customer and produce the necessary amount of parts to obtain higher income.

The part numbers consist of kits, each kit can contain up to 18 pieces, and orders are between 20 and 60 kits. The unit price of the high voltage cables is approximately 15 dollars, and to know the value of the order, the unit value of the cable must be multiplied by the number of pieces per kit by the number of kits requested in the order. It is expected to produce 800 pieces per day, but with the current production process, only 432 pieces are produced.

Currently, the company generates $157,427.44 US dollars (USD) from sales, so the profit per piece is $364.41. However, the company is not gaining the expected profit due to process failures, as shown in Fig. 4.6.

It has been detected that in the production process in cell 2311, the production part number XYZ, specifically in the workstations, the following problems occur:

- Bottlenecks.
- Delay in production, resulting in late delivery to the customer.
- High overtime payment to workers regularly.
- Unnecessary movements when executing operations.
- High production cost.
- High defective parts (121 parts/day on average).

From these problems detected in the production cell, the workers and the company generally are the most affected (besides the final customer). All of the above for the company represents a problem since it has to pay overtime when required, representing more economic losses than profits. In the same way, regarding costs, not having the standardized production cell represents an increase in production cost per piece since the longer it takes for the piece to leave the process, its cost increases. In addition to the above, not making deliveries on time generates a bad organizational image for the customer due to failure to meet deadlines.

High voltage cables are manufactured in cell 2311, and the space where the production process is carried out (i.e., the physical location) lacks adequate equipment and furniture and does not have an optimal layout. Currently, the cell does not have a shelf or a defined space to place the cables, which causes disorder and mixing of raw material because it is not identified and separated according to different production orders and part numbers, giving a bad image to the cell, since it is not possible to identify the flow process. Figure 4.7 shows the current condition of the cell due to the lack of an established place for the cables.

As seen, the cables accumulate and are not identified with the part number and corresponding order, which confuses the operators and mixes the material. It is difficult to

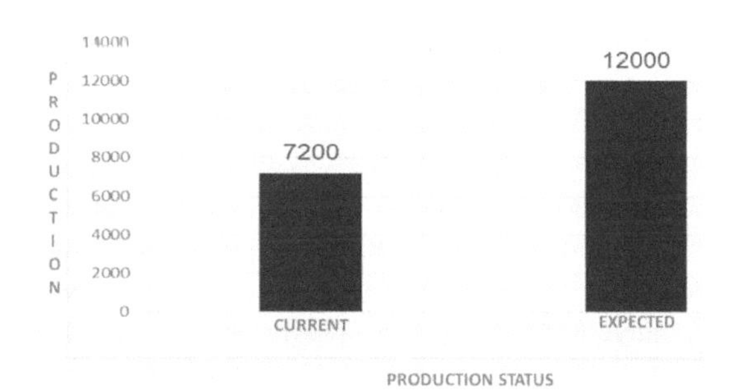

Fig. 4.6 Current and expected state of daily production value in dollars

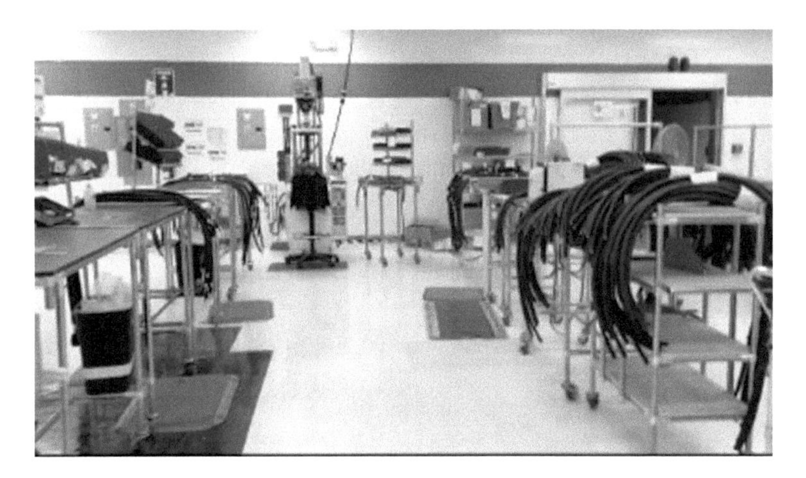

Fig. 4.7 Current condition of the area

know which work orders each cable corresponds to, and here waste is identified, which is the overprocessing since by not identifying the material, a measurement of each of the pieces must be made to know which part number it corresponds to, and this results in idle time. This task is tiring for the operators, which causes that sometimes they do not do it in the right way, and mixes have produced that lead to rework and loss of raw material.

Another issue within the cell is the poor distribution area. There is a material rack where the boxes sent by the Warehouse Department are placed with the raw material needed to carry out the order, such as terminals and bags. This rack is located outside the cell, and this causes a relatively long distance to walk to get the raw material and supply it to the workstations; this leads to another waste as transportation.

Unnecessary trips are also a loss for the company because it translates into downtime. Another aspect where the poor layout can be identified is in the process flow; that is, the cell is designed in a U-shape, which causes a waste of space, as shown above in Fig. 4.5.

In addition, high voltage cable cell 2311 currently has obsolete equipment for carrying out the process. The cable stripping operation is carried out in the process, and since it is a relatively large gauge, it is not easy to operate completely manually, so there is a semiautomatic stripping machine that facilitates this task. The current stripping machine is very old (see Fig. 4.8), so it no longer performs its function correctly and causes damage to the cable, which in turn causes the pieces to be rejected.

As can be seen, the stripping machine is damaged, in addition to being outdated equipment, since there are now digital strippers that facilitate the operation. Also, riveting is carried out in the manufacturing process of high-voltage cables and fixtures are used to hold the cable so that it does not move during the impact of the terminal placement. The problem with this operation is that the fixtures do not hold the cable properly, so the cable moves; in addition, they are not attached to the riveting machines, they are only attached

Fig. 4.8 Semiautomatic obsolete stripping machine

with double adhesive tape, so it is not possible to have total confidence at the time of riveting. Figure 4.9 shows the fixture.

For the labeling operation, templates are used in which the cables that will be produced and the designator that each of them will carry can be seen graphically. The problem is that a template must be used for each part number, even if all the part numbers have the label placed at the same distance, which causes a loss of time in searching the templates. Even if the distance and procedure are the same for each model, it is impossible to run the production line without the template requested by the PI (process instruction), and it causes idle time in each model change, even though it is the same operation.

According to those above, it can be detected that different wastes do not add value to the product within the high voltage cable work cell.

Fig. 4.9 Current rivetting structure

4.2.2 Research Objectives

4.2.2.1 General Objective

The main objective of this case study is to increase the daily production rate of the XYZ assembly in cell 2311 to meet the expected production.

4.2.2.2 Specific Objectives

The specific objectives of the project are the following:

- Define the manufacturing process flow for high voltage cables.
- Reduce high-voltage cable waste by at least 50%.
- Increase the company's profits by at least 25%.

4.3 Methodology

After defining the project and mentioning the opportunity areas to work on and the objectives to be achieved, the methodology is developed, and the PDCA cycle and other tools are used to solve the current problems.

Next paragraphs are mentioning how each of them is applied to the project's development. There are the following stages:

1. Stage 1. Plan
2. Stage 2. Do
3. Stage 3. Check
4. Stage 4. Act.

4.3.1 Stage 1. Plan

This first stage consists of 4 steps, which are described below.

1. Define the problem: As already mentioned throughout the document, high voltage cables are a product that benefits the company economically due to the high income they provide. To define the problem, detection of the different defects present in these cables, as well as other wastes present in the manufacturing process, is carried out. Table 4.1 shows a Project Charter, which is used to identify more clearly the problem to be solved, the objectives to be achieved with the project and the people involved.

Table 4.1 Project charter

	Project title:				
	Project leader:	Project start date:	Project end date:	Product:	
Description of the process		Problem statement			
Project goal		Scope			
Benefits		Measurable project objectives		Resources and support required	
Approval/steering committee		Stakeholders and advisors		Project team and SMEs	
Name	Organization	Name	Organization	Name	Organization

The Project Charter shows the problem to be solved and the project's main objectives. Another tool that visualizes the project's scope and the people directly affected by the product is the suppliers, inputs, process, outputs and customers (SIPOC) diagram.

According to Yeung [13], the SIPOC diagram is a systematic tool to monitor products and services for customer satisfaction as it helps to build a link between workers, products, workplace, price, promotion, and customer needs satisfaction. According to the American Society of Quality (ASQ), the SIPOC diagram is a tool for collecting data on all relevant elements of a process improvement project before starting work [13, 14]. Figure 4.10 presents the SIPOC diagram used in this project.

To identify the customer requirements and what he expects from the company, the voice of the customer (VOC) diagram is used [15]. From this diagram, it is possible to know what aspects to work on. Figure 4.11 shows an example of a voice of the customer diagram.

2. Describe the current state: To have a more in-depth concept of the current manufacturing process of high-voltage cables, a flow chart is developed to break down each of the phases and decisions are taken within the manufacturing process. A flowchart is a graphic representation that breaks down a process in any activity developed in industrial or service companies and their departments, sections or areas of their organizational structure Serrano-Cobos [16]. These diagrams are important as they help to designate any graphical representation of a procedure or part of it. Nowadays,

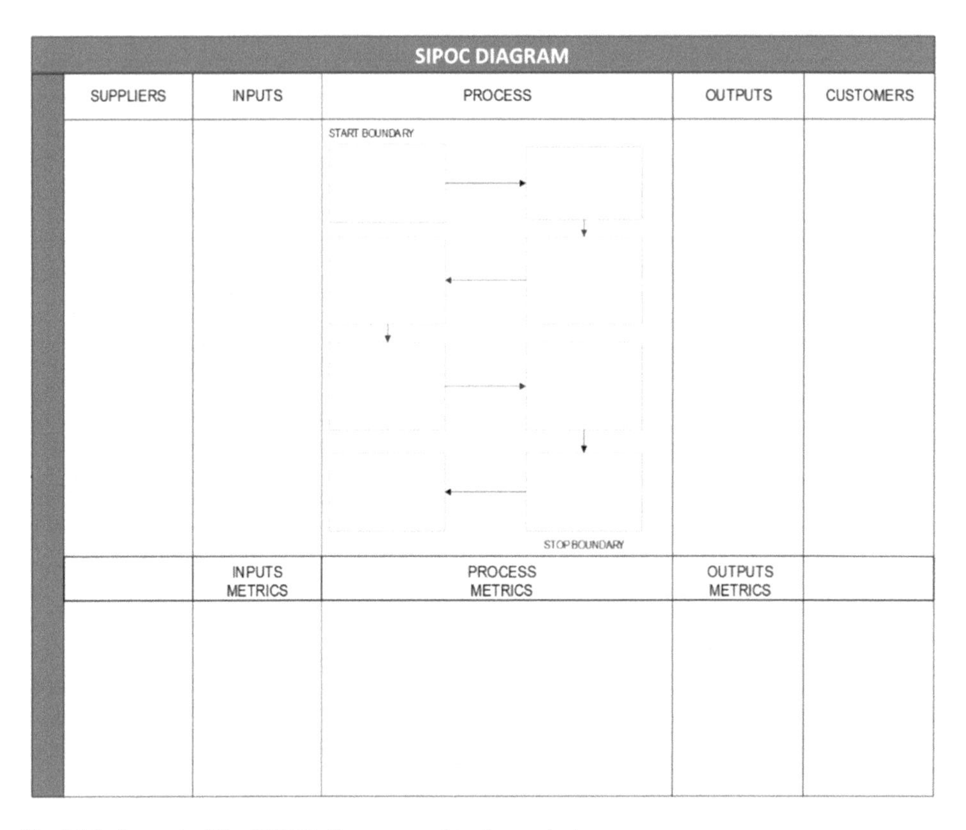

Fig. 4.10 Layout of the SIPOC diagram used in the project

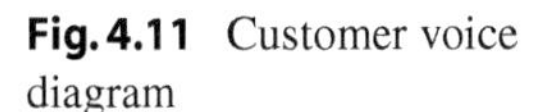

Fig. 4.11 Customer voice diagram

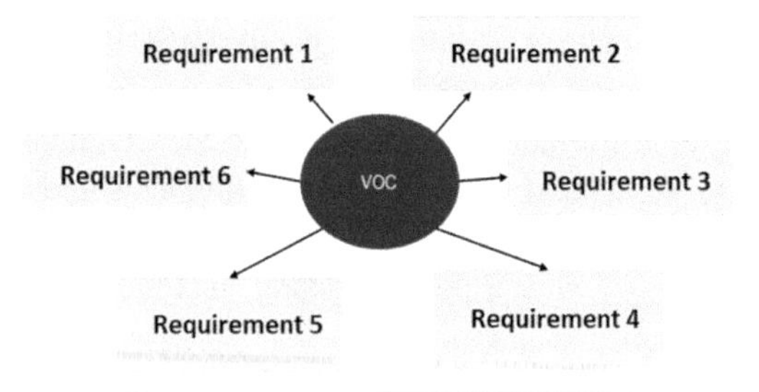

flowcharts are considered in most companies as one of the main instruments in the realization of any method or system [16].

Within the company, a format is used to keep track of parts produced hour by hour, the quality rejects that occur with the parts produced in each hour, and idle time. This format is called a control sheet or log sheet. The control sheets are structured forms that facilitate information collection, previously designed based on the needs and characteristics of the data required to measure and evaluate one or more processes. To calculate idle time per hour, the formula presented in Eq. (4.1) is used:

$$\text{Idle time} = \frac{\text{Produced pieces} \times \text{expected cycle time-3600 s}}{60\ \text{min/h}} \tag{4.1}$$

As mentioned above, the expected cycle time is 53 seconds/part. As an example of Eq. (4.1), if 37 parts were produced during one hour, the idle time calculation would give the following result:

$$\text{Idle time} = \frac{37\ \text{pieces} \times 53\ \text{s/piece-3600 s}}{60\ \text{min}} = 27.31 minutes/hour$$

During the data collection process, a series of videos are taken in which the operators who currently work in the cell are interviewed to know the problem firsthand since they are the ones who face the obstacles that exist every day.

Each operator comments on improvements and areas of opportunity that need to be worked on.

Considering the operators' contributions is necessary because this allows them to have a more comfortable and pleasant workplace. To better understand the area's current situation, a wiring diagram is made to record the route that the cables run in the cell. In this way, it will be possible to observe the transport and the wasted space currently available.

3. Identify the root causes: Once the current state of the cell and the problems that occur every day are described, we look for potential root causes that generate these problems, as well as possible solutions to improve the situation in the area. The Ishikawa diagram is used to observe those causes graphically. Subsequently, we proceed to identify the causes that are most relevant to the problem, that is, to identify the root causes of the effect.

   A cause-effect matrix is used as a tool, where a weighting is assigned to each cause concerning the different criteria: quality, customer service, safety, and production, where the level of relevance and impact governs the assigned weighting. Once each cause is rated, it is averaged to obtain the final value. The following scale is used for the magnitude of the cause:

   1–2: Not appreciated.

   3–4: Appreciated, but low.

   5–6: Mitigation measures need to be analyzed and considered.

   7: This may mean project development conflicts and require more detailed analysis or studies.

   Similarly, the following nomenclature is used to assign values to the causes in the cause-effect matrix:

   - Significant Negative Impact—RED
   - Medium negative impact or warning of possible significant impact—YELLOW

- Irrelevant impact—GREEN.

According to the cause-effect matrix and the average results obtained, a Pareto diagram is made to graphically identify and observe the root causes of the greatest impact. The Pareto diagram helps to define the causes to which priority should be given to solve the problem and avoid wasting resources by eliminating causes that are not of great relevance. Pareto's law, also known as the 80/20 rule, states that in general and for many phenomena, approximately 80% of the consequences come from 20% of the causes [17]. A Pareto diagram is a special bar chart where the plotted values are arranged from highest to lowest [18]. In this project, a Pareto diagram is used to identify the most frequently occurring defects, the most common causes of defects, or the most frequent causes of customer complaints.

4. Solution and implementation plan: After having identified the root causes of the problem, we proceed to select the activities and tools that will provide a solution to the problem. A Gantt chart is used to describe the activities that will be carried out to reduce and eradicate the problem, in addition to selecting the people in charge of carrying out these activities and specifying the deadline for their completion. The Gantt chart is a graphic tool that aims to expose the time of dedication foreseen for different tasks or activities along a determined time. With this diagram is possible to know if the improvements are implemented in time and form.

4.3.2 Stage 2. Do

This is the core phase of the methodology since this is where the improvements to solve the problem are implemented, and the Gantt chart is used. These activities help to define a new manufacturing process and support the cell with the necessary resources for efficient production.

4.3.3 Stage 3. Check

After implementing the solutions, we proceed to observe the cell's behavior and determine whether the improvements made are correct and help improve the situation in cell 2311. An internal company report is presented, documenting the activities carried out and their impact on the area. Table 4.2 shows the format used for the internal report.

The report describes each of the improvements made in the cell; this report is the format used within the company to document each of the projects carried out and have evidence of what is implemented, in addition to observing the benefits obtained and analyzing the results. In order to keep track of the high voltage cable cell and verify that the improvements implemented can be maintained and that the results improve considerably,

Table 4.2 Internal company report format

2020 Productivity improvement project						
Objective:			Management owner:			
Team:						
Current situation summary						
Obstacle	Action	Who?	What do you expect?	Results	Meet expectations	Experience (Describe what you learned when you applied the action)

Table 4.3 Internal company report format

Cell:	Leader:		Date:
Check list for improvement control			
Points to evaluate	Yes	No	Observations

a checklist is used to verify that the improvements are ongoing and that the time and resources invested are rewarded. Table 4.3 shows the checklist.

With the checklist, control of the improvements is kept since the format is filled out daily to ensure that they are being fulfilled correctly. In addition, a record is kept verifying the validity and operation of the improvements over time. This checklist also allows the operators to contribute ideas on the areas of improvement that they identify and informs them of the aspects that need to be worked on.

4.3.4 Step 4. Act

In this phase, we analyze whether the implemented improvements are the right ones, i.e., we reflect on what we have learned. In addition, the next step for the continuous improvement of the project is specified.

4.4 Results

This section presents the results obtained by implementing the method described in the previous section. For a better understanding, these results are presented for each phase of the method described.

4.4.1 Findings from Phase 1: Plan

Figure 4.12 shows the result of the SIPOC diagram of the JCI high voltage cable manufacturing process. According to the diagram, the process and metrics were not rigorous, so the expected results were not obtained. Therefore, the production process was modified, and metrics were established to help obtain the necessary production and quality.

SIPOC DIAGRAM				
SUPPLIERS	INPUTS	PROCESS	OUTPUTS	CUSTOMERS
• Cutting area • Warehouse • Label and documentation supplier	• Cable • Terminals • Labels • IP	START BOUNDARY Material arrival → Assortment of material at workstations → Cable marking with adhesive tape → Riveted → Inspection → Labeled → Packaging → Inspection by Quality Department STOP BOUNDARY	• High voltage cables of various lengths and gauges	
	INPUTS METRICS	PROCESS METRICS	OUTPUTS METRICS	
	• Know the process • Supply of material at workstations • Prepare work area	• Receive material at each workstation • Know the process • Place the correct number of cables in each kit	• Correct packaging • Cable in good condition	

Fig. 4.12 SIPOC diagram of the project

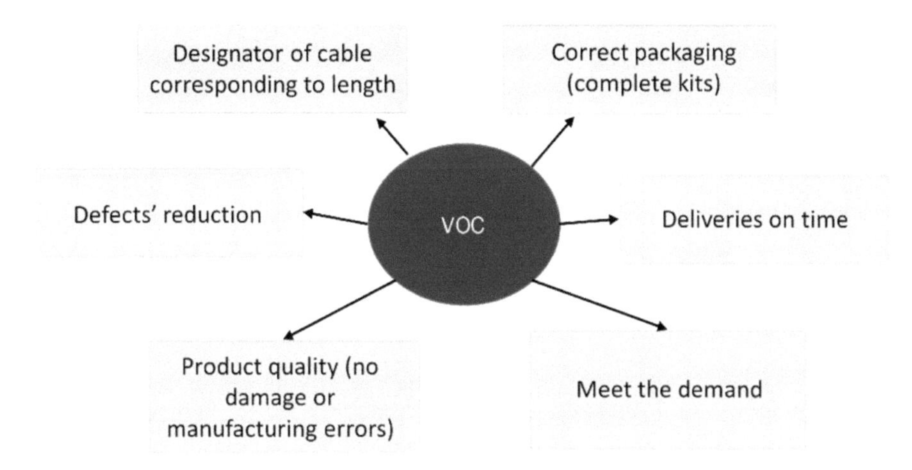

Fig. 4.13 Result of the VOC diagram

On the other hand, Fig. 4.13 shows the VOC diagram in which it is observed that the customer wants quality and on-time delivery, so the project is compatible with the customer's requirements.

Similarly, Fig. 4.14 shows the result of the initial flow diagram of the manufacturing process of the high-voltage cables. In this case, it can be seen that there was a repeated operation within the process since, in the beginning, the cable was identified by placing a piece of adhesive tape on it, and at the end of the process, it was removed, and a label with the same designation was placed, which generated waste of unnecessary operations (over-processing).

On the other hand, regarding the initial production, in the control sheet shown in Fig. 4.15, you can appreciate the daily production in cell 2311 on September 17, 2019, where we worked with the part number XYZ. At the end of the day, 432 parts were produced. However, 121 parts were rejected for quality due to rivet, length and part damage. The idle time were 296 min.

As can be seen in Fig. 4.16, previously, there was considerable transportation in terms of obtaining the raw material, since considering the orders variety and the constant changes of models, it generated a loss of time in the journey, and there was wasted space since the previous design of the cell did not allow to use the maximum space of the work area.

Figure 4.17 shows the detected causes of the low production level and the defects in the produced parts. In contrast, Table 4.4 shows the weightings and the average of these obtained for each of the causes. Figure 4.18 shows the Pareto diagram of the project with the main causes to which the greatest effort and resources to improve the situation of the cell by increasing the production and decreasing the defects.

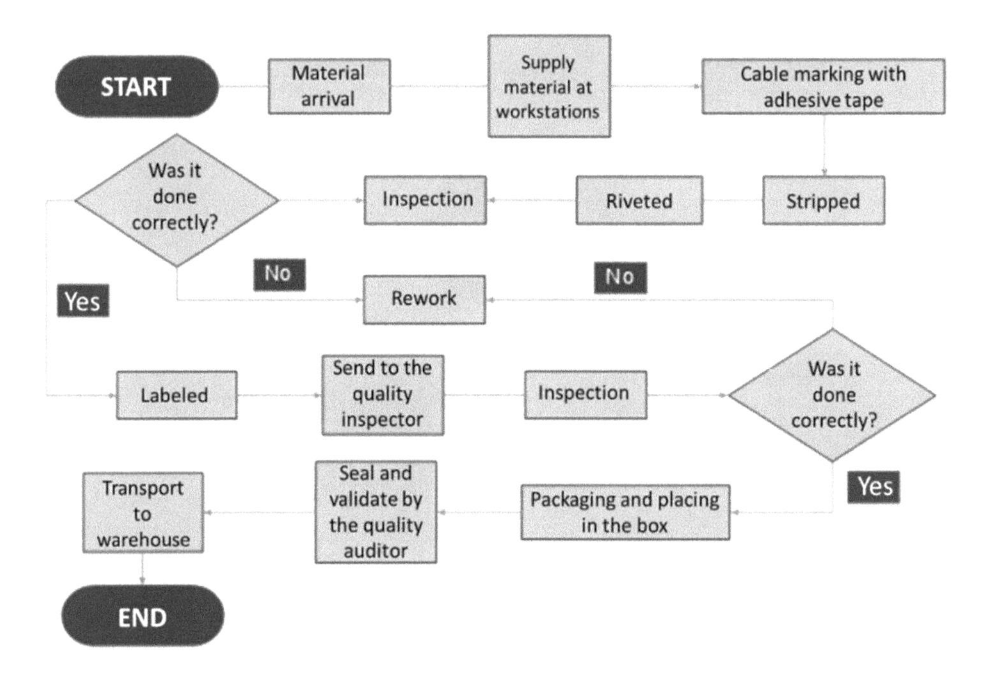

Fig. 4.14 Initial flow diagram of the high-voltage cable manufacturing process

According to the results obtained in the Pareto diagram, it can be identified that the main causes that must be solved to improve the problem are:

- Poor manufacturing process
- Wrong process flow
- Complicated process instructions to understand
- Unnecessary operations
- Mixing of material.

Therefore, these causes turned out to be the ones that we have an interest in solving to improve the cell's situation.

4.4.2 Findings from Phase 2: Do

The first causes to which solutions were found were the following:

- Poor manufacturing process
- Wrong process flow
- Unnecessary operations.

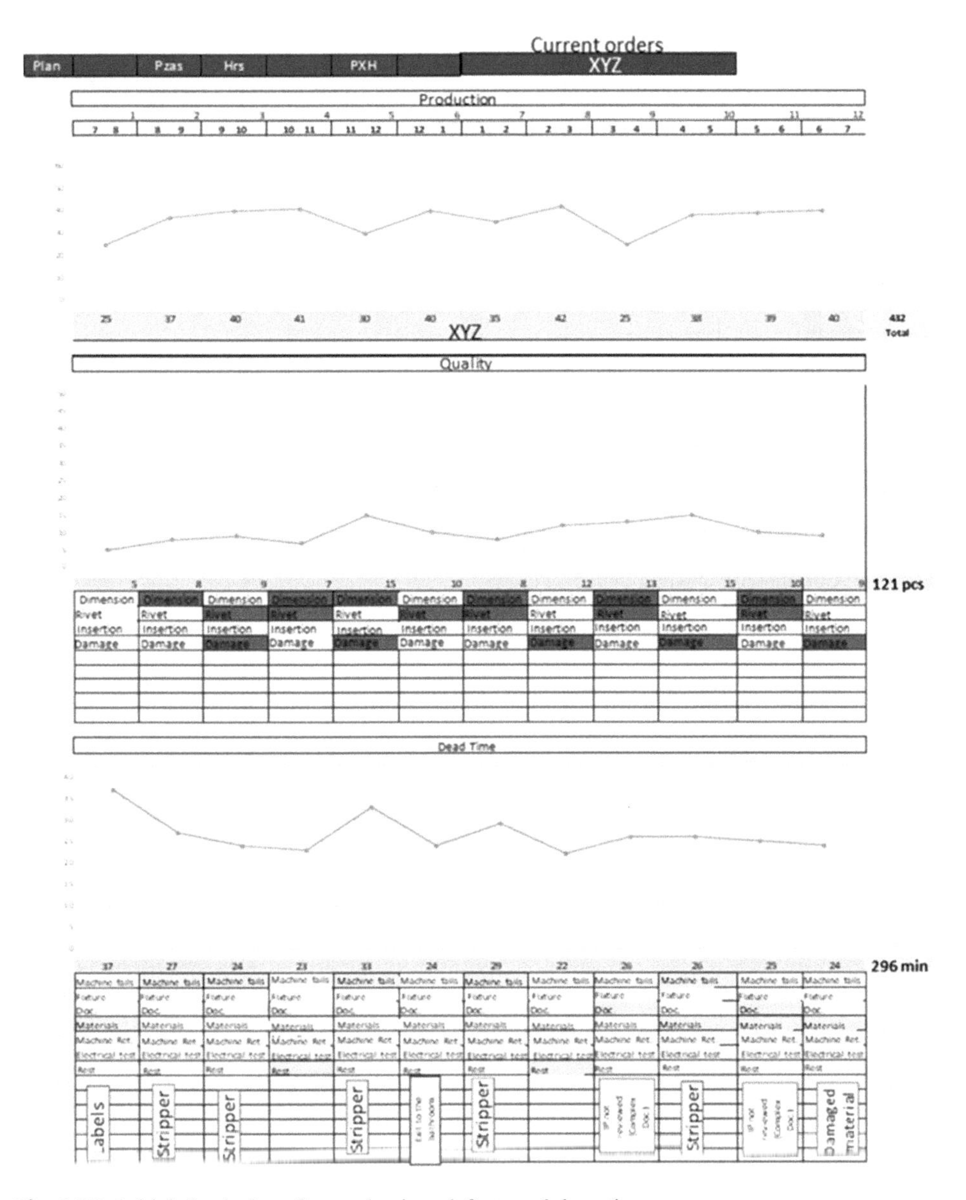

Fig. 4.15 Initial check sheet for production, defects and downtime

These causes are those that are directly affecting the manufacturing process. With this, the production equipment was renewed, and the work area and environment for the cell operators were improved. In addition, unnecessary operations were eliminated, and a new, more efficient production flow was designed. Figure 4.19 shows the flow chart for the cell improvement where the manufacturing process of high voltage cables is restructured. On

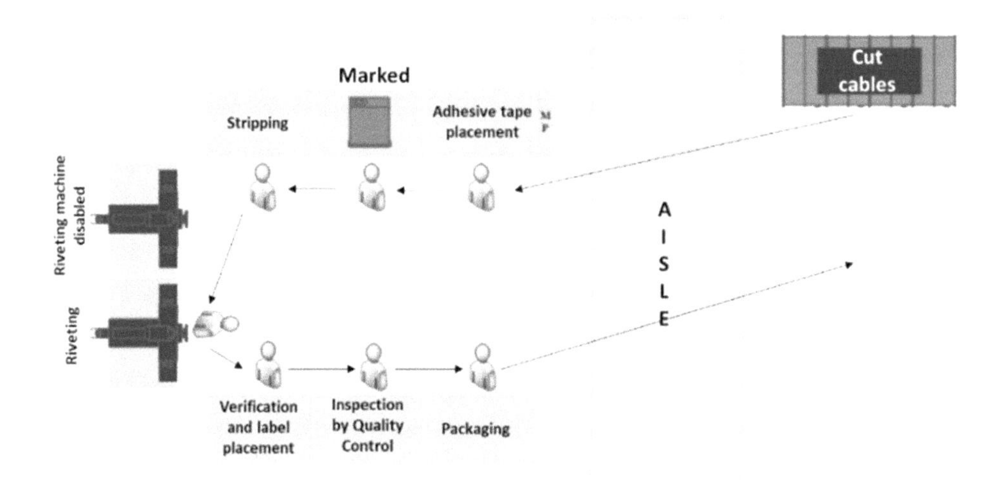

Fig. 4.16 Current wire diagram of cell 2311

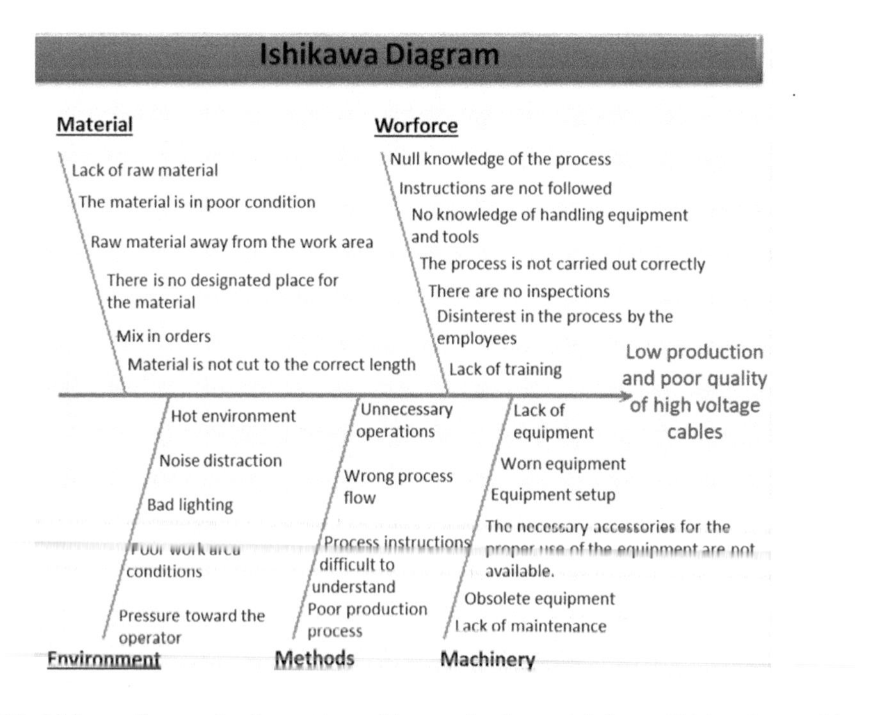

Fig. 4.17 Ishikawa diagram for the problem of low production and defects of high-voltage cables

the other hand, Table 4.5 shows the updated process diagram, where the improvements implemented to increase production and reduce quality defects can be seen.

As seen in the updated process diagram, the operation of marking with adhesive tape was eliminated since it was an unnecessary and repetitive operation. To eliminate this operation, a first piece validation was implemented; that is, the piece is measured and verified to have the length specified in the visual aids and drawing. Based on its length and caliber characteristics, the corresponding designator is placed. This way, it is ensured from the beginning that the piece goes correctly, thus avoiding repeating operations and reducing the cycle time of the piece. It can also be seen in the new manufacturing process that the relabeling operation was replaced by the final inspection, where it is verified that the terminals and the orientation of the rivets are correct based on customer specifications, reducing inspection time by the quality inspector and increasing the release of parts and sent to the customer.

On the other hand, one of the problems detected in the Pareto diagram, which was solved, was the mixing of material. This problem was because there was no designated

Table 4.4 Cause-effect matrix

Number	Causes	Criteria				
		Quality	Customer Service	Security	Production	Average
1	Lack of equipment	4	5	1	7	4.25
2	Obsolete equipment	4	5	1	7	4.25
3	The necessary accessories for the proper use of the equipment are not available	7	4	6	7	6
4	Equipment setup	2	2	1	3	2
5	Worn equipment	4	2	2	6	3.5
6	Lack of maintenance	2	2	2	2	2
7	Null knowledge of the process	6	6	6	3	5.25
8	Instructions are not followed	7	6	6	5	6
9	No knowledge of handling equipment and tools	6	6	6	4	5.5
10	Lack of training	6	7	6	5	6
11	The process is not carried out correctly	6	7	5	6	6
12	Disinterest in the process by the employees	5	3	1	5	3.5
13	There are no inspections	7	6	1	6	5
14	Lack of raw material	2	7	1	5	3.75
15	The material is in poor condition	7	7	1	6	5.25
16	Raw material away from the work area	2	1	1	7	2.75
17	There is no designated place for the material	3	1	1	7	3
18	Mixing of material	7	7	4	7	6.25
19	Material is not cut to the correct length	7	2	1	3	3.25
20	Poor manufacturing process	7	7	4	7	6.25
21	Complicated process instructions to understand	7	7	4	7	6.25
22	Wrong process flow	7	7	4	7	6.25
23	Unnecessary operations	7	6	5	7	6.25
24	Pressure toward the operator	1	1	4	2	2
25	Poor work area conditions	1	1	5	5	3
26	Bad lighting	1	1	1	1	1
27	Noise distraction	1	1	1	1	1
28	Hot environment	1	1	1	1	1

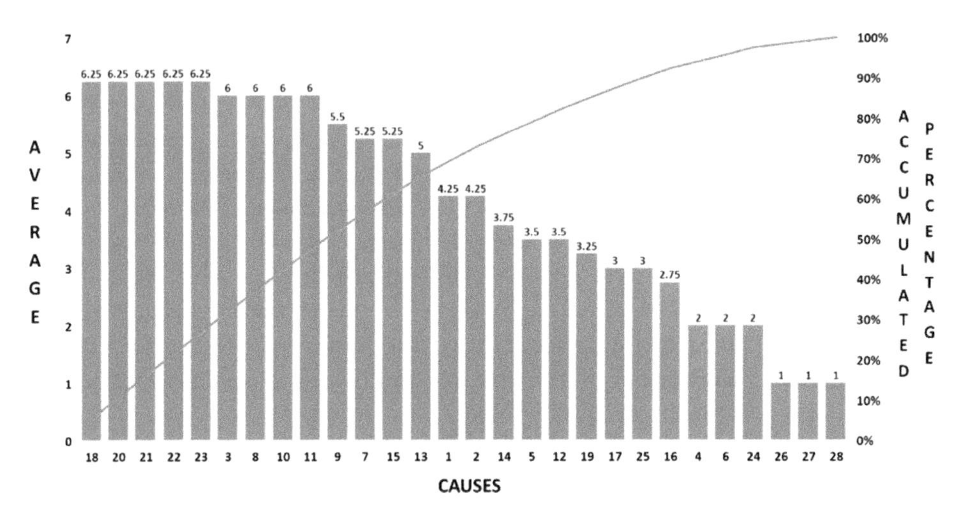

Fig. 4.18 Pareto diagram for the causes of the problem in cell 2311

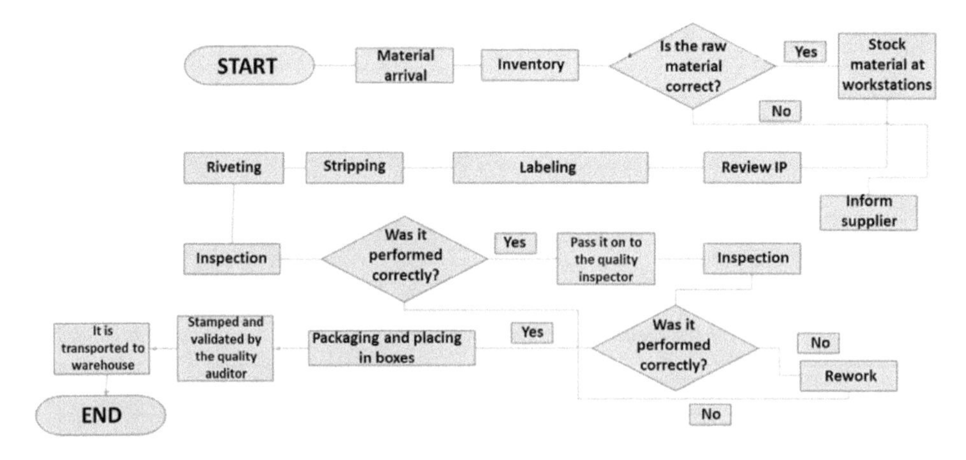

Fig. 4.19 Improved flow diagram

place for the placement of the orders to be run, and they were not identified or separated from each other. It is important to have a material rack with separators and identifiers to solve this problem. To this end, an internal search was carried out in the plant to recycle the material that was not being used. During the search, a rack of material ready to be used was discovered in the Hold Material area. The rack did not have identifiers, so the work team started to make them with the help of the Corel program, a printer, and a laminating machine.

Table 4.5 Improved flow process chart

No	Operation	●	➡	D	▼	■	Remarks
1	The arrival of raw material	●					Ensure that the material is complete according to the order and defects-free
2	Raw material inventory and inspection	●					
3	Tagged	●					
4	First part validation				■		Check that the part matches the drawing and IP
5	Semi-stripping	●					
6	Stripping	●					
7	Riveting	●					
8	Part inspection				▼		Check that the final piece is free of defects
9	QC inspection				■		Validate final parts and pack
10	Material release by QC auditor				■		Release material boxes with QC stamps
Stock transfer to warehouse			➡				Send parts to the customer
Summary	6	1	0	0	4		

Figure 4.20 shows the material racks implemented in the cell, with their corresponding separators and the material identifiers created by the equipment.

It can be seen that a rack of material was placed for the cables and another for the rest of the material, which are terminals, IP and labels. On each rack were placed separators and material identifiers to reduce the mixing of orders and have a place assigned for each thing, obtaining a better order in the cell as a benefit. Similarly, an identifier was placed on each bin, including the part number, quantity and order to which they belong. In addition, racks of material were also implemented to place the finished product, ready to be inspected for quality.

Additionally, a second riveting machine was enabled. This was done in order to reduce riveting time and increase production. With this, each person riveted one side of the cable and thus improved the flow. Originally, the riveting operation was the bottleneck in the manufacturing process, so it was decided to install a second riveting machine already in the plant but not used for maintenance purposes. Figure 4.21 shows the two riveting machines in the cell.

Fig. 4.20 Implemented material rack

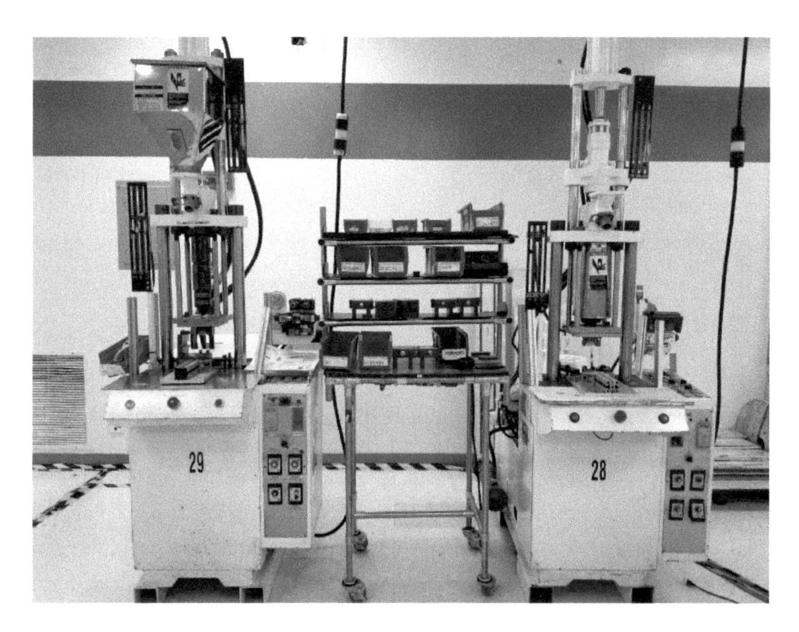

Fig. 4.21 Existing riveting machines in cell 2311

As can be seen, now there are two riveting machines and a rack where the bins with the terminals to be used are placed, according to the production order and the process instructions. As previously mentioned, the riveting machine had a fixture for the placement of the cable during the impact of the rivet, but this fixture was not properly attached to the machine, which caused failures when riveting because the piece was not held properly and moved, so new fixtures were created, which proved to be safer and facilitated the operation. The final result of the new fixtures is shown in Fig. 4.22.

Fig. 4.22 New riveting fixtures

As you can see, an aluminum base was created that is screwed to the riveting machine, giving greater security and better support. The fixtures are placed on this base, and the cable is on top. The fixtures were designed as a press so that, once the cable is placed, it closes and avoids any movement during the riveting process. The aluminum base was designed so that the fixtures are interchangeable and will fit any gauge fixture. This implementation improved the quality of the parts by reducing poorly riveted parts and increasing operator safety since the fastener does not open, and the cable is not dropped.

Previously it was mentioned that to perform the stripping operation, there was an obsolete stripping machine that did not mark the piece correctly, so it was decided to implement a new updated and intelligent stripping machine, that is, with a higher technological level. The stripper implemented in the cell had already been purchased by a work team that had previously tried to increase the production of the cell without obtaining results. Because of this, the current work team dedicated themselves to studying the manual and introducing it into the cell. Figure 4.23 shows the new stripper.

Fig. 4.23 New stripper

As can be seen, the new stripper has a higher technological level. In this equipment, it was possible to program the Setup for each gauge and type of insulation, which resulted in less time for model changes since it was not necessary to adjust the stripping machine from start to finish, but only to place the option stored in the memory that corresponded to the characteristics of the cable. In addition, one worker was eliminated in this operation, since before, it was necessary to have one worker marking the piece and another removing the insolation. On the other hand, this machine has better blades, which allowed a better marking, and it was easier to remove the heatstroke, even being able to remove the heatstroke completely in certain cases.

Another aspect that was decided to change within the cell was the labeling templates. These templates were used to place the designator on the cable at a certain size. The problem with the templates is that they had to be changed whenever a model change was made since, according to the IP specifications, the one that was made especially for that model had to be used. However, the placement of the labels on all part numbers is 2 inches from the tip, and the only difference is the length of the cable.

Therefore, it was decided to standardize a labeling method and implement a fixture where the cable is placed, which already has the measurement of inches and is self-adjusting to the size of the cable to be labeled. The fixture was made with the help of the tool room equipment, so the implementation was inexpensive and manufactured in a short period of time. Figure 4.24 shows the new labeling fixture.

As you can see, the cable is placed inside the fixture, each end of the cable goes into a fixture, and the label is placed from where the fixture ends, as it was designed exactly 2 inches apart. One of the fixtures is fixed to the table, and the other has a screw at the bottom to adjust or readjust the fixture to move it to the required length. With this fixture, the downtime caused by searching for the templates at each model change was

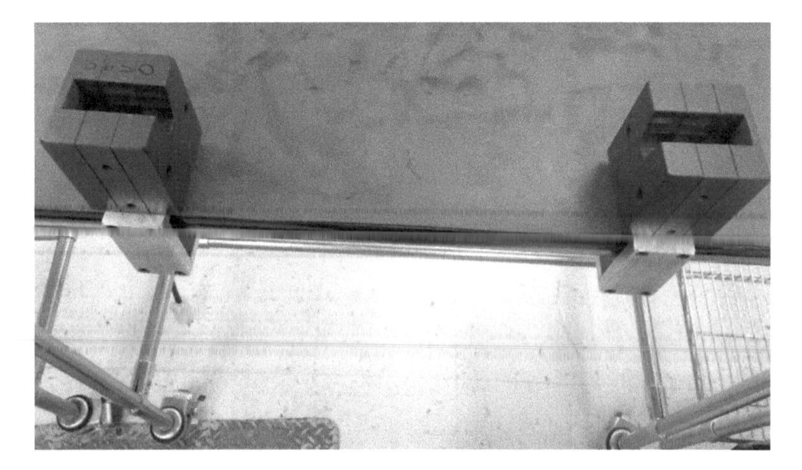

Fig. 4.24 New labeling structure

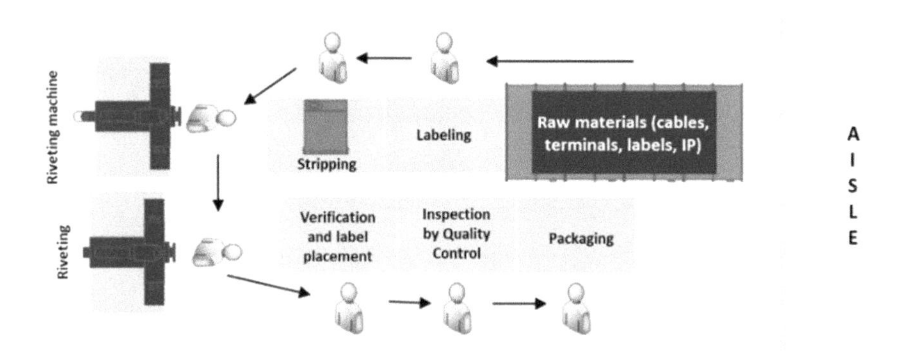

Fig. 4.25 Improved plant layout

eliminated, thus standardizing the labeling operation. On the other hand, the templates became obsolete, and the wood used to place the templates could be reused better and provide benefits in other areas.

Previously it was commented that the layout was not adequate for the cell since the space was not used to the maximum and long distances were traveled to acquire the material, so it was decided to optimize the space and change the design of the cell to the mirror form so that in this way the space would not be wasted and other cells could be placed. Figure 4.25 shows the new floor layout.

As can be seen, the new distribution of the area maximizes space and has a continuous flow of the piece, in addition to reducing the transport distance to acquire the raw material, as it is now in the cell and it is not necessary to move or travel long distances. With this option, the space between stations was reduced. Finally, a solution was given to the last of the causes that were determined as important after having carried out the Pareto: process instructions that were complicated to understand and to improve it, two activities were carried out, both did not mean any cost for the company, since they were carried out with the resources that existed in the plant.

The first solution was to make samples of the figures of the rivets; that is, to make a wire to scale where the orientation of the blow and the figure to be made are clearly shown. With these samples, it was easier for the operators to understand the riveting since they had a tangible piece of how it should be done, and they could have it at the workstation without any problem consulting it at any time. Figure 4.26 shows the samples.

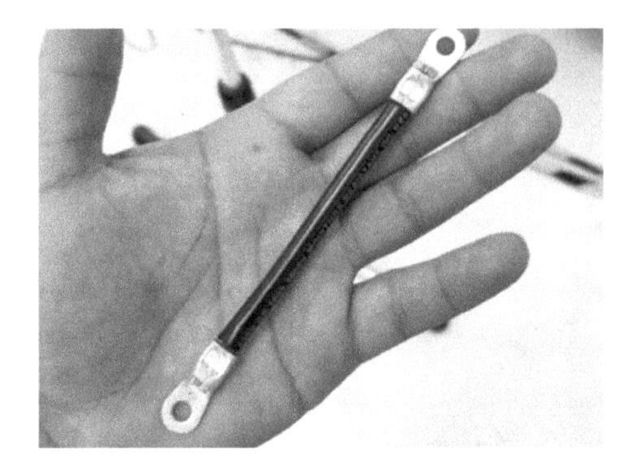

Fig. 4.26 Riveting sample

As can be seen, the sample proved to be a very useful tool, as it is practical and indicates the orientation for terminals and rivet strike. With this solution, the defects due to riveting errors were considerably reduced. The second solution was to update the visual aids to the improved format in the company since it is more graphic and the indications are better appreciated. Figure 4.27 shows one of the new visual aids.

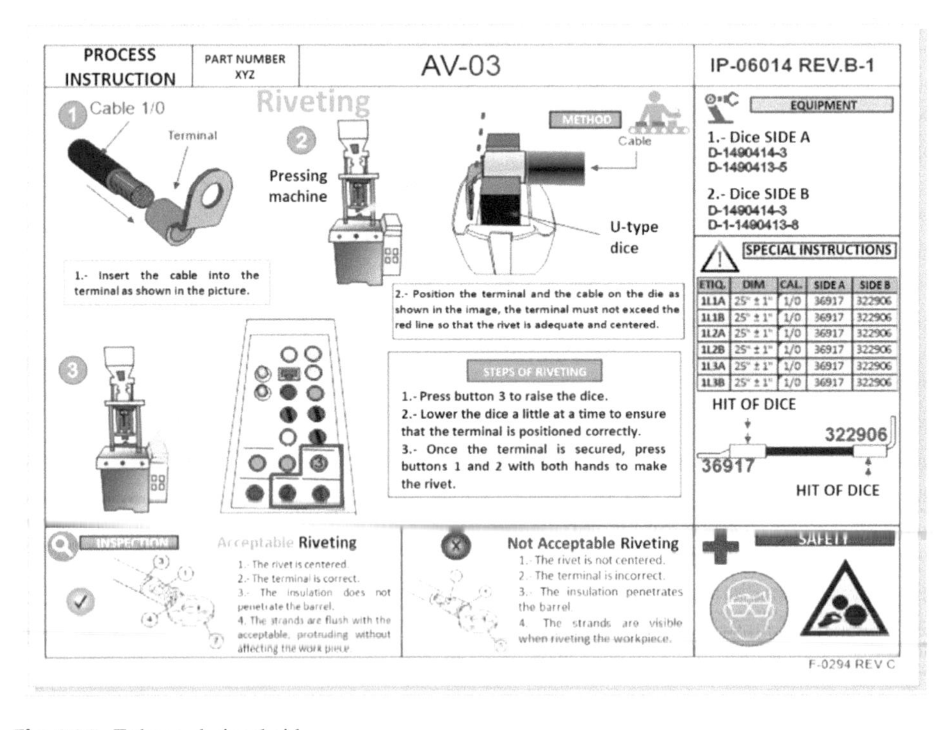

PROCESS INSTRUCTION | PART NUMBER XYZ | AV-03 | IP-06014 REV.B-1

Riveting

1 Cable 1/0

Terminal

1.- Insert the cable into the terminal as shown in the picture.

2 Pressing machine

METHOD

Cable

U-type dice

2.- Position the terminal and the cable on the die as shown in the image, the terminal must not exceed the red line so that the rivet is adequate and centered.

3

STEPS OF RIVETING

1.- Press button 3 to raise the dice.
2.- Lower the dice a little at a time to ensure that the terminal is positioned correctly.
3.- Once the terminal is secured, press buttons 1 and 2 with both hands to make the rivet.

EQUIPMENT

1.- Dice SIDE A
D-1490414-3
D-1490413-5

2.- Dice SIDE B
D-1490414-3
D-1-1490413-8

SPECIAL INSTRUCTIONS

ETIQ.	DIM	CAL.	SIDE A	SIDE B
1L1A	25" ± 1"	1/0	36917	322906
1L1B	25" ± 1"	1/0	36917	322906
1L2A	25" ± 1"	1/0	36917	322906
1L2B	25" ± 1"	1/0	36917	322906
1L3A	25" ± 1"	1/0	36917	322906
1L3B	25" ± 1"	1/0	36917	322906

HIT OF DICE

322906

36917

HIT OF DICE

INSPECTION

Acceptable Riveting

1.- The rivet is centered.
2.- The terminal is correct.
3.- The insulation does not penetrate the barrel.
4. The strands are flush with the acceptable, protruding without affecting the work piece.

Not Acceptable Riveting

1.- The rivet is not centered.
2.- The terminal is incorrect.
3.- The insulation penetrates the barrel.
4. The strands are visible when riveting the workpiece.

SAFETY

F-0294 REV C

Fig. 4.27 Enhanced visual aid

As can be seen, the visual aids were more graphic and understandable, which allowed for fewer operator errors. This format is friendlier to people and has all the necessary indications.

4.4.3 Findings from Phase 3: Check

The control sheet shown in Fig. 4.28 shows that, in addition to increasing production from 432 to 813 parts/day (an increase of 88.2%), there was also a decrease in quality defects (from 121 to 31 defective parts). However, there are still damages in the parts, which correspond to an external cause to the cell, since the terminals are supplied in the cell with aesthetic damages, such as exposed base metal, which is a defect that does not pass the quality criteria. Finally, the dead time was reduced from 296 to 24 min.

As can be seen, the visual aids were more graphic and understandable, which allowed for fewer operator errors. This format is friendlier to people and has all the necessary indications.

4.4.4 Findings from Phase 3: Check

The control sheet shown in Fig. 4.28 shows that, in addition to increasing production from 432 to 813 parts/day (an increase of 88.2%), there was also a decrease in quality defects (from 121 to 31 defective parts). However, there are still damages in the parts, which correspond to an external cause to the cell, since the terminals are supplied in the cell with aesthetic damages, such as exposed base metal, which is a defect that does not pass the quality criteria. Finally, the dead time was reduced from 296 to 24 min.

4.4.5 Findings from Phase 4: Act

The implemented improvements were the correct ones since benefits were obtained, and currently, it is not necessary to make corrections to the project since it is stable and has constant improvement. However, it was detected that it is important to improve the visual signaling in the workplace; that is to say, to place in the cell and workstations a kind of traffic light that indicates the status of the process in which the cell is, so an Andon with different colors was placed. The colors would be defined as follows:

- Green: Producing
- Red: Detained
- Yellow: QC Inspection
- Blue: Model change.

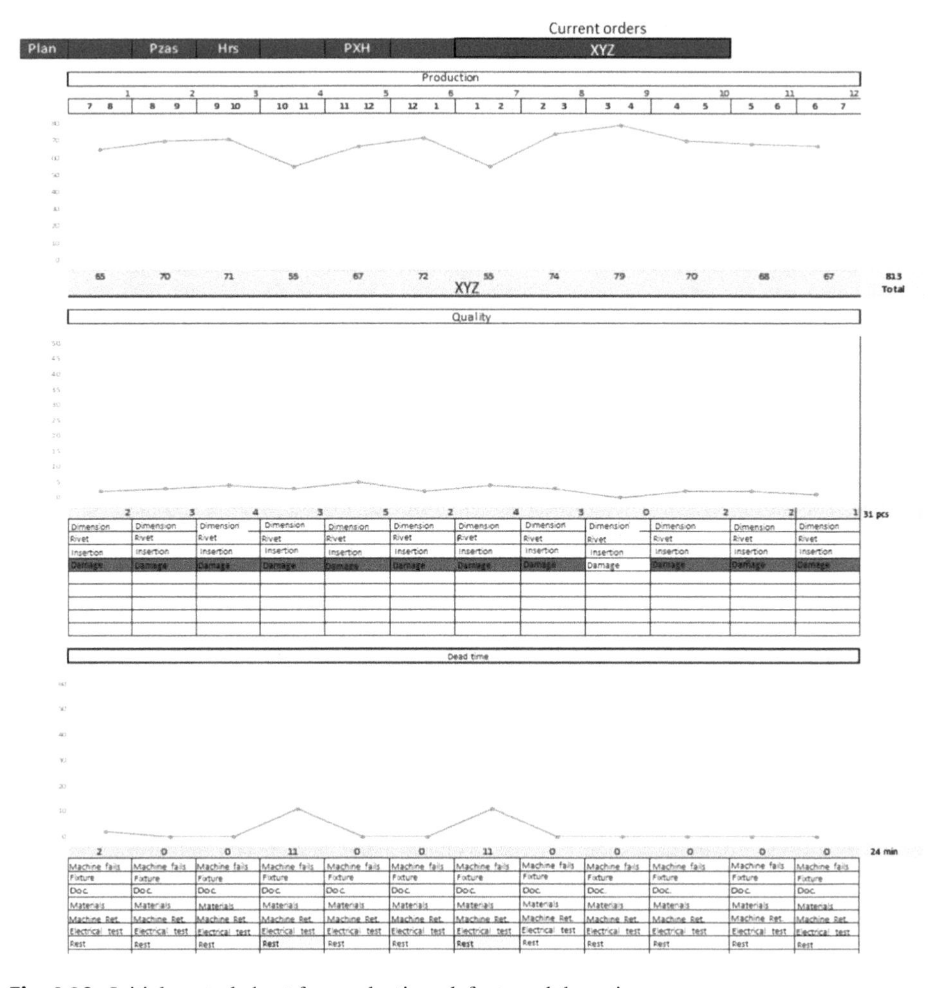

Fig. 4.28 Initial control sheet for production, defects and downtime

This walkway allowed anyone passing through the cell to know what production phase it was in, giving greater standardization to the area and visual management. On the other hand, these traffic Andon also served to inform the status of the material in the work stations, which avoided stopping the cell for lack of material and informed the supplier when to provide the raw material. The nomenclature of the colors of the workstation platform was as follows:

- Green: Stocked
- Red: Material missing
- Yellow: Little material left

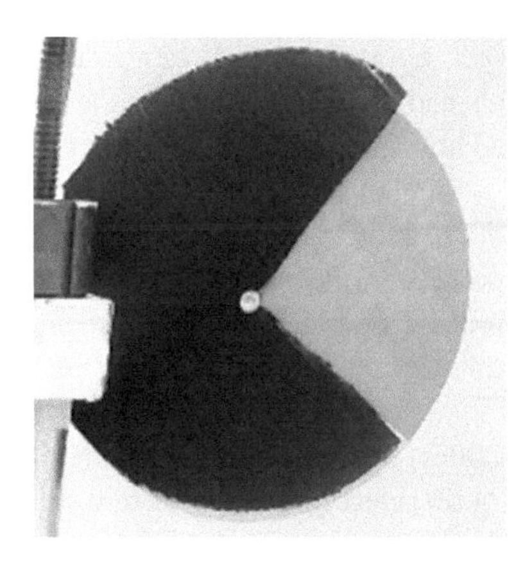

Fig. 4.29 Andon system implemented in cell 2311

- Blue: Model change.

Figure 4.29 shows a reference image of the walkers implemented in cell 2311.

4.5 Conclusions

The project's overall objective was to increase the daily production of high voltage cables to meet the demand. According to the results shown above, a change in the daily production level can be appreciated, where it is demonstrated that, after the implemented methodology, this objective was met since it went from producing, on average, 432 to 813 pieces/day. This result agrees with the one obtained by Silva et al. [19], who applied the PDCA cycle to reduce the Can Loss Index (CLI); that is, to increase an organization's production. As a result, the annual CLI decreased from 0.97% to 0.78% in the first year, reaching a value of 0.60% in the second year. In a similar case study, they applied the PDCA cycle to determine the main causes that affect the decrease in machine production in a clothing company with socks products and to provide solutions to increase production. As a result, the investigation yielded an increase in production output of 112% per machine/month.

Regarding the specific objectives, the first one was to define the manufacturing process flow for high voltage cables. From this objective, it can be concluded that it was achieved satisfactorily, since at the beginning (see Fig. 4.16) there was a longer route to transport the cut cables to the workstations, as well as the transport of these between stations. With the new distribution of the cell, a continuous flow of material was achieved, and unnecessary operations were eliminated (placement of adhesive tape).

In addition, the new process flow reduced transportation waste, unnecessary movements, over-processing and waiting. This result is consistent with that of Ikatrianasari and Putra [10], who applied PDCA in a multinational company in the connector and terminal industry and obtained; as a result, a reduction in the time spent on transporting materials and products, as well as the percentage of defective products by 82.6% (defects). In addition, 51 product inspection activities (over-processing) were eliminated.

Similarly, Nino et al. [12] implemented the PDCA cycle in a Rural Hospital where there were transport wastes, waiting time, unnecessary movements, defects and overprocessing. As a result, these authors could reduce/eliminate these wastes, making the process flow more stable, and unsterilized trays were no longer sent to the operating room by mistake or trays were no longer sterilized twice.

The second specific objective was to reduce at least 50% of the defects in high voltage cables. According to the results obtained with the control sheets, defects decreased by 74.5%, since previously there were up to 121 defective parts per day, and after the project, there were only 31 defective parts per day. Therefore, it is concluded that this second objective was achieved satisfactorily.

This finding is similar to that obtained by Jagusiak-Kocik [20], who conducted a case study on the practical use of the PDCA cycle in a manufacturing company in the plastics processing industry in the small and medium-sized enterprise sector. This author applied the PDCA cycle to solve quality problems in manufacturing photo frames: discolorations and burns on the frame surface. When measures were introduced to reduce the number of defects, a decrease of more than 60% was observed.

In another similar research, Nugroho et al. [21] applied the PDCA cycle to examine and analyze the factors that cause the occurrence of waviness and curvature defects in the coil spring plate. As a result of their research, these authors reduced the frequency of occurrence of such defects.

Finally, the third specific objective was to increase the company's profits by at least 25%. As it is known, previously, the company generated profits of $157,427.44. With the increase in production to 813 pieces, the profit increased to $296,269.70, representing an increase of 88.2%. Therefore, it can be concluded that the third specific objective was successfully achieved.

This result is similar to that obtained by Silva et al. [19], who, by applying the PDCA cycle, managed to reduce the CLI, also reducing costs (and therefore increasing profits) by more than 35% in the first year and by 28.91% in the first four months of the second year. In another similar case, Zadry and Darwin [22] applied the cycle in a company dedicated to manufacturing handmade shoes. In that company, problems related to defects in footwear products were detected at an average of 12% per month. After applying the PDCA cycle, defects were reduced to 0%, while losses due to defective footwear products were reduced from Rs. 1,250,000 per week to zero. This enabled the company to increase its profit from sales.

In general conclusion, the PDCA cycle, supported by tools such as the Ishikawa diagram and Pareto diagram, allows the elimination of different wastes, such as defects, overprocessing and transportation, which translates into higher production and increased profits due to sales.

References

1. Gupta S, Jain SK (2013) A literature review of lean manufacturing. Int J Manag Sci Eng Manag [Internet] [cited 2022 Jun 29]; 8(4):241–249. https://www.tandfonline.com/doi/abs/10.1080/17509653.2013.825074
2. El-Namrouty KA, Abushaaban MS (2013) Seven wastes elimination targeted by lean manufacturing case study "gaza strip manufacturing firms". Int J Econ Financ Manag Sci [Internet] [cited 2019 Feb 19] 1(2):68–80. http://www.sciencepublishinggroup.com/j/ijefm
3. Furman J, Malysa T (2021) The use of lean manufacturing (LM) tools in the field of production organization in the metallurgical industry. Metalurgija [Internet] [cited 2022 Jun 29] 60(3–4):431–433. https://hrcak.srce.hr/clanak/372293
4. Wahab ANA, Mukhtar M, Sulaiman R (2013) A conceptual model of lean manufacturing dimensions. Procedia Technol 1(11):1292–1298
5. Capital M (2004) Introduction to lean manufacturing for Vietnam. Mekong Capital Ltd
6. Rawabdeh IA (2005) A model for the assessment of waste in job shop environments. Int J Oper Prod Manag 25(8):800–822
7. Shah D, Patel P (2018) Productivity improvement by implementing lean manufacturing tools in manufacturing industry. Int Res J Eng Technol [Internet] [cited 2018 Oct 3] 5(3):3794–3798. www.irjet.net
8. Dixit A, Dave V, Singh AP (2015) Lean manufacturing: an approach for waste elimination. Int J Eng Res Technol. 4(4):532–536
9. Jangid A (2019) Implementation and analysis of true lean in a startup company by using PDCA model, a case study in a manufacturing venture [Internet]. University of Kentucky. https://uknowledge.uky.edu/ms_etds/9/
10. Ikatrianasari EAPH, Putra ZF (2012). Penerapan lean manufacturing melalui Metode Gemba Kaizen dengan Pendekatan Siklus PDCA untuk Peningkatan Produktivitas di PT. XYZ, Bekasi. In: SNTI III-2012. Jakarta, Indonesia: SNTI III-2012 Universitas Trisakti, pp I014-1–I014-6
11. Nurafiqah MZ (2019) A study on the adoption of lean management techniques. Case study. In: Bakery Industry [Internet]. Melaka; [cited 2022 Jul 5]. https://digitalcollection.utem.edu.my/24903/
12. Nino V, Claudio D, Valladares L, Harris S (2020) An enhanced Kaizen event in a sterile processing department of a rural hospital: a case study. Int J Environ Res Public Heal [Internet] [cited 2022 Jul 5] 17(23):8748. https://www.mdpi.com/1660-4601/17/23/8748/htm
13. Yeung S-C (2008) Using six sigma-Sipoc for customer satisfaction. In: Antony J (ed) 3rd international conference on six sigma: lean six sigma as a vehicle for a successful business transformation. UNIVERSITY OF STRATHCLYDE, Glasgow, Scotland, pp 359–379
14. ASQ (2022) SIPOC+CM Diagram [Internet]. SIPOC+CM Diagram [cited 2022 Jul 8]. https://asq.org/quality-resources/sipoc
15. Rese A, Sänn A, Homfeldt F (2015) Customer integration and voice-of-customer methods in the German automotive industry. Int J Automot Technol Manag 15(1):1–19
16. Serrano-Cobos MR (2019) Optimización de la cadena logística [Internet], 1st ed. Spain: Elearning, 208 pp. https://books.google.com.mx/books?id=C3flDwAAQBAJ&pg=PA21&dq=Un+diagrama+de+flujo+es+una+representación+gráfica+que+desglosa+un+proceso+en+cualquier+tipo+de+actividad&hl=es&sa=X&ved=2ahUKEwi9z7iL-O74AhW4JkQIHR_1CB0Q6AF6BAgIEAI#v=onepage&q=Undiagramadef
17. Lipovetsky S (2009) Pareto 80/20 law: derivation via random partitioning. [Internet] [cited 2022 Jul 10] 40(2):271–277. https://www.tandfonline.com/doi/abs/10.1080/00207390802213609

18. Joiner Associates I (1995) Pareto charts: plain and simple [Internet]. In: Reynard S (ed) Oriel incorporated. Madison, 130 pp. https://books.google.com.mx/books?id=Mubz8xTERqEC&printsec=frontcover&dq=Pareto+Charts:+Plain+%26+Simple&hl=es&sa=X&ved=0ahUKEwjU3tmZ2fzdAhWNLHwKHRxvDw8Q6AEIJzAA#v=onepage&q=ParetoCharts%3APlain%26Simple&f=false
19. Silva AS, Medeiros CF, Vieira RK (2017) Cleaner production and PDCA cycle: practical application for reducing the Cans Loss Index in a beverage company. J Clean Prod [Internet] [cited 2018 Oct 10] 150:324–338. https://www.sciencedirect.com/science/article/pii/S0959652617304687
20. Jagusiak-Kocik M (2017) PDCA cycle as a part of continuous improvement in the production company - a case study. Prod Eng Arch [Internet] [cited 2018 Oct 29] 14:19–22. https://yadda.icm.edu.pl/baztech/element/bwmeta1.element.baztech-d32f115f-abac-4be1-a931-84a85bf48ab3;jsessionid=FB6F0E724880E8283B33FF1B2DB5BCBE
21. Nugroho RE, Marwanto A, Hasibuan S (2017) Reduce product defect in stainless steel production using yield management method and PDCA. Int J New Technol Res. 13(11):39–46
22. Zadry HR, Darwin R (2020) The success of 5S and PDCA implementation in increasing the productivity of an SME in West Sumatra. In: IOP conference series: materials science and engineering [Internet]. IOP Publishing, Medan, Indonesia [cited 2022 Jul 17], pp 1–10. https://iopscience.iop.org/article/10.1088/1757-899X/1003/1/012075